VOLUME TWO

# DISCRETE SEMICONDUCTORS AND OPTOELECTRONICS

BY SY LEVINE

Printed in the United States of America

Published by ELECTRO-HORIZONS PUBLICATIONS

114 Lincoln Road East
Plainview, New York 11803

# ACKNOWLEDGEMENT

My very special thanks to Jerry Worthing, my good friend and mentor, and to Esther, my devoted and loving wife, without whose dedicated, discerning and meticulous editing efforts this volume would not have been completed. Their suggestions, advice and unending "manuscript honing" were inestimably helpful in the final production of this book. To them, I extend my deepest gratitude and affection for their tireless toil and trouble.

S.L.

# TABLE OF CONTENTS

# PART II - OPTOELECTRONICS

# PREFACE

This series of books is written for the men and women in industry, education and the military whose need for a conceptual comprehension of electronics has become increasingly urgent. These books are dedicated to those who feel the impact of the technology and are motivated to explore its complexities, to obtain a working knowledge of electronics and to improve their technical communicative skills. The books are intended for those people in purchasing, sales, marketing, production, advertising, management and electronics classes who will find this information particularly beneficial in their work.

BASIC CONCEPTS AND PASSIVE COMPONENTS, the first volume of the 3-volume set, A LIBRARY ON BASIC ELEC-TRONICS, provided a study of basic electronic concepts, terminology, its rules and contents, while offering insights into electronic technology. Passive components were examined as to how they are made, how they work, how they are specified on a data sheet and how they relate to each other in electronic circuitry and systems.

This second volume in the series, DISCRETE SEMICONDUC-TORS AND OPTOELECTRONICS, continues the study by examining the active components of circuitry. In a step-by-step progression, more of the totality of electronic circuitry is developed to provide a more comprehensive picture of this technology and state-of-the-art developments.

As with Volume One, the information in this book is enhanced with specific reinforcement exercises (and answers), glossaries and a definitive index for easy reference.

# PART ONE

# DISCRETE SEMICONDUCTORS

# INTRODUCTION TO SEMICONDUCTORS

During the latter part of the Nineteenth Century, relatively unsophisticated electrical circuits were used in the lighting, heating, telephone and telegraph systems that were available at the time. There were no radios, televisions, audio amplifiers, communications systems or computers, because the technology required for their design did not yet exist. Electrical *amplification* was the catalyst needed to move the existing technology into the era of *electronics*.

Amplification is the process of increasing or intensifying an electrical signal without changing its appearance or shape. Because the devices needed to provide amplification had not yet been developed, electronics was still a science of the future. It was the work of two scientists that pushed amplification closer to becoming a reality.

In 1903, Sir John Fleming, a British scientist, invented a two-element device that he called a *valve*. It later became known as a *vacuum tube diode*, the forerunner of the *vacuum tube triode*, developed in 1907 by Lee DeForest, an American inventor. The vacuum tube triode was a three-element device with the unique capability of amplifying electrical signals when used as part of a circuit. Because of this unique feature, the science of Electronics came into being.

By the early 1920s, vacuum tubes, the amplifying components in electronic circuits, opened the way to the development of radio receivers, radio transmitters, audio amplifiers, sophisticated instrumentation systems and similar equipment. By the 1930s, radio became the major medium for electronic communications, and television, still in its infancy, was a laboratory novelty.

Vacuum tubes generally served the industry well, despite having several disadvantages.

- Because of their size and design, vacuum tubes occupied large areas, used considerable power and required a large power supply that needed additional space.
- Part of their structure consisted of filaments that generated a great amount of heat within a circuit. Because of its destructive effect, the heat had to be removed.
- Vacuum tubes usually lasted approximately two years before burning out. Preventive maintenance procedures required periodic replacement before burn-out.
- Vacuum tubes were not shock or vibration-proof.

Many other components used in a vacuum tube circuit also consumed large amounts of power and generated additional heat, the "enemy" of electronic system reliability. If heat was not removed from a system's enclosure, the values of sensitive elements in its circuit could change, possibly causing circuit failure. It was necessary to use cooling equipment that removed heat but added to the overall weight and size of a system. A different approach was crucial if the drawbacks inherent in the use of vacuum tubes were to be eliminated.

In 1938, a new research program was initiated at the Bell Telephone Laboratories in New Jersey to explore and utilize the electrical characteristics of solid materials, particularly germanium. Before this time, other types of solid materials, such as galena, copper-oxide and selenium had been studied, but had limited use. Compared to the vacuum tube, these materials were called *solid-state* because their properties allowed electrical conduction to occur within their solid crystalline mass, whereas electrical conduction that occurred within the vacuum tube was across its open structured elements.

At the start of World War II, this work was temporarily diverted to other programs but was resumed in 1946. In 1948, three Bell Telephone Laboratories scientists, John Bardeen, Walter Brattain and William Shockley announced their invention of the *transistor* and in 1956 received the Nobel Prize in Physics for their efforts. The transistor was destined to replace the vacuum tube—the component that had been the heart of electronic circuitry for a half-century and was soon to become obsolete.

This new device, the transistor, became one of a group of the solid-state devices that were called *semiconductors*. The Semiconductor Industry began in 1954, ushering in an era of electronic innovation. In a relatively short time, general acceptance of this new and expanding technology created an unprecedented outpouring of a variety of semiconductors. The transistor, one of the more popular semiconductors, caught the imagination of the electronics industry and "transistorization" became the new byword in the electronics vocabulary.

Many related industries emerged in support of the semiconductor manufacturing effort that required the use of device packages, diffusion ovens, automatic test equipment and the supplies and facilities needed for clean-room operation. Since semiconductor technology is based on photography, metallurgy and chemistry, the research and development efforts in these sciences also expanded rapidly.

New manufacturing techniques helped to lower production costs, improve yields and lower prices. Many companies were able to expand their product line and modify older types of equipment to attain better performance features at lower costs. With the use of semiconductor circuitry, new state-of-the-art electronic equipment was made possible.

Because of their unique characteristics, semiconductors have many advantages over vacuum tubes for use as circuit elements.

- System power is appreciably reduced.

- Since decreased power results in decreased heat, the need for heat transfer equipment, such as fans, blowers and other cooling devices is greatly reduced.

- Space and weight requirements in a system are diminished, opening the way to microminiaturization in electronic equipment packaging.

- Improved circuit performance is achievable.

- Development of new types of portable, battery-powered products are made possible.

- Since the working element in these new devices is made of one solid piece of material, they are shock and vibration-proof. They can be launched, dropped, vibrated and mechanically agitated with no detrimental effects.

- Since semiconductors have no volatile filaments, essentially infinite longevity can be achieved, resulting in greatly improved system reliability.

Because semiconductors provided substantial reduction in weight, size, power and improved equipment reliability, the space industry became a reality. The equipment required for communications, navigation and instrumentation had to be light in weight, compact and use a minimum of power. These features were impossible with the use of vacuum tube circuitry.

The tiny semiconductor was used in the development of a highly advanced level of medical instrumentation. Obviously, it would be extremely impractical to implant a pacemaker made with vacuum tube circuits into a human body. The small, extremely low-power, highly reliable semiconductors were the essential components of the electronic circuits used in the complex systems that operated artificial limbs and other highly sophisticated prosthetic devices.

Semiconductor technology made possible an enormous diversity of new consumer products. These included extremely small portable radios, miniature TV sets, hand-held and desk-top calculators, digital watches and a vast variety of video toys. In diverse industries such as automotive, rotary equipment, industrial control and combustion engineering, bulky electromechanical controls have been replaced by low-power, space-saving, maintenance-free semiconductor circuits.

In 1954, when the semiconductor industry began, the only available computers were massive mainframes that used vacuum tube circuitry, occupying large areas of space while consuming huge amounts of power. Because of semiconductor technology, the same types of computers, as well as the newer minicomputers and microcomputers, now provide superior performance, use considerably less power, are more reliable and often can be housed in small cabinets "no bigger than a bread box."

In comparing semiconductors with vacuum tubes, consideration of their reliability is of utmost importance. Reliability consists of two elements—*quality*, a defined degree of excellence, and *longevity*, the usable life-span. Although a vacuum tube may have had outstanding quality, its working elements would wear out in a relatively short time. By the standards of the period, the vacuum tube was considered a "reliable" device, but if subjected to present-day requirements, it would be unacceptable.

Semiconductor reliability also depends on its quality and longevity. Product quality can be achieved through controlled manufacturing techniques and established quality control standards at the production facility. Longevity of semiconductors, however, is determined by factors other than quality alone. Their rated voltage, current, power and environmental specifications must not be exceeded when functioning in a finished system. If a quality product is produced and the longevity factors are met, in all likelihood, semiconductors would never need to be replaced.

Like any product where quality is critical, the quality of a semiconductor is a prime requirement, however, its longevity has become the basic factor in its universal acceptance and success.

Semiconductors have established themselves as a major influence in the astounding expansion of electronics technology during the last three decades. Since the era of the vacuum tube, no other components have so dramatically changed the nature and direction of the electronics industry and have so profoundly influenced all other industries throughout the world.

 CHAPTER
ONE

# DEFINITIONS AND GENERAL INFORMATION

THE FAMILY OF DISCRETE
SEMICONDUCTORS

MATERIALS USED TO PRODUCE
SEMICONDUCTORS

THE SEMICONDUCTOR
MECHANISM

THE ELECTRONIC SWITCH

SUMMARY

# DEFINITIONS AND GENERAL INFORMATION

A *discrete semiconductor* is a single, solid-state, active component enclosed in a finished package. It has the capability of differentiating between a positive and negative voltage and can function as a conductor, as a nonconductor, or as a voltage-controlled resistor.

The generic name, "semiconductor," refers to each of the unique components produced by solid-state technology. These components are called semiconductors since any device so designated is capable of functioning as a conductor and as a nonconductor. Its specific mode of operation depends on the amount of applied voltage, its polarity and the manner in which it is connected in a circuit. Because the prefix "semi" means half, the term "semiconductor" is actually misleading. A semiconductor has the unique characteristic of having several functions, since it has conducting and nonconducting capability as well as the ability to act as a fixed or variable resistor.

## THE FAMILY OF DISCRETE SEMICONDUCTORS

There are several major subdivisions within the "family" of discrete semiconductors and they are categorized in Figure 1.

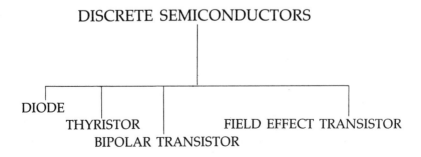

**Figure 1**

| DESIGNATION (NAME) | APPLICATION |
|---|---|

## DIODE

| | |
|---|---|
| General purpose diode | Switch, Demodulator |
| Computer switching diode | High-speed switch |
| Photodiode | Changes light to electricity |
| **Rectifier/Bridge rectifier** | Changes A.C. to pulsating D.C. |
| Zener diode | Voltage regulator<br>Voltage reference<br>Circuit protection (clamp) |
| Light emitting diode (LED) | Indicator light<br>Digital display |

## THYRISTOR (A.C. SWITCH)

| | |
|---|---|
| Silicon Controlled Rectifier | Power control circuits<br>Transient voltage protection |
| TRIAC | Lighting control<br>Heating control |

## BIPOLAR TRANSISTOR

Classified by structure:
NPN (Negative/Positive/Negative material) or
PNP (Positive/Negative/Positive material)

| | |
|---|---|
| General purpose and audio bipolar transistor | Low-frequency amplifier<br>Low-speed digital switch |
| Radio frequency and high-speed bipolar transistor | High-frequency amplifier<br>High-speed digital switch |

| DESIGNATION (NAME) | APPLICATION |
|---|---|

## FIELD EFFECT TRANSISTOR (FET)

Classified by structure:
N-channel, P-channel

| | |
|---|---|
| Junction FET (J-FET) | General purpose amplifier<br>Audio amplifier<br>Voltage-controlled resistor |
| Metal Oxide Semiconductor FET<br>(MOS FET) | R.F. amplifier<br>Digital and Analog switch |
| Complementary MOS (CMOS) FET | Digital and Analog switch |

Discrete semiconductors are further classified as:

POWER COMPONENTS
Devices capable of dissipating 3 watts or more at 25°C.

SMALL-SIGNAL COMPONENTS
Devices capable of dissipating less than 3 watts at 25°C.

MILITARY OR AEROSPACE COMPONENTS
These devices conform to military specification MIL-S-19500. They are specified to operate from -55°C to +125°C and to be stored in an atmosphere ranging from -55°C to +150°C. They are encased in hermetically-sealed glass or metal packages. This subject is covered in CHAPTER SIX — SEMICONDUCTOR RELIABILITY CONSIDERATIONS.

CONSUMER, INDUSTRIAL AND
COMMERCIAL COMPONENTS
There is no standard operating temperature range for these devices. Their rated operating range is specified on a data sheet for each type. They are normally enclosed in plastic packages and typically operate from 0°C to +70°C.

# MATERIALS USED TO PRODUCE SEMICONDUCTORS

| MATERIAL | SYMBOL | END PRODUCTS | OPERATING TEMPERATURE |
|---|---|---|---|
| Selenium | Se | Rectifiers | 0°C to + 85°C |
| Germanium | Ge | Computer switching diodes & high frequency transistors | -55°C to + 100°C |
| Silicon | Si | Diodes, rectifiers, zeners, transistors, SCRs, TRIACs, FETs, I/Cs (Linear & digital) | -55°C to +150°C |
| Gallium Arsenide | GaAs | LED indicators & LED digital displays, | |
| Gallium Arsenide Phosphide | GaAsP | Opto-couplers & Solid-State Relays | -55°C to + 100°C |
| Gallium Phosphide | GaP | High frequency diodes & transistors | |

Basic Chemical Elements Used In Producing Semiconductors
**Figure 2**

## SELENIUM

Selenium (Se) was one of the first chemical elements used in the commercial production of semiconductors. These components are made by depositing a thin layer of nonconductive selenium onto one surface of an aluminum plate and then coating the selenium with a conductive metal. This structure allows the free flow of current from the selenium layer to the conductive coating beneath it, however, current is prevented from flowing in the opposite direction. By its one-way action, the device is made to perform as a *rectifier* and acts to change A.C. into pulsating D.C.

The operating temperature of selenium ranges from 0°C to +85°C. Components made from this material are limited to commercial, industrial and consumer applications. One unique feature of a selenium rectifier is the distinctive odor given off by the material when it burns out. The defective component can quickly be isolated "by nose" and promptly replaced, simplifying the process of troubleshooting.

# GERMANIUM

The element germanium (Ge), in wafer form, is processed by diffusing arsenic and indium, in gaseous form, into its structure. The diffusion process changes the metallurgical nature of germanium into a material used to manufacture semiconductors. An inherent feature of germanium is the fast movement of its molecules, or the *high mobility* within its structure. This high mobility provides the fast switching characteristics needed for diodes used in computer circuits. It also affords the high frequency response needed for those transistors used in radio frequency amplifiers.

The operating temperature range of germanium (-55°C to +100°C) only covers industrial, commercial and consumer applications. When changes in temperature occur, significant shifts are produced in the electrical characteristics of germanium semiconductors, often resulting in circuit instability. Although germanium is still used as the base element for some semiconductors, it has largely been replaced by silicon as the preferred material for the manufacture of semiconductors.

# SILICON

In the late 1950s, silicon (Si) became the major element used in the production of semiconductors. In its natural state, silicon is found in ordinary beach sand in the form of silicon dioxide. Next to air and water, sand is the most plentiful compound on earth. Next to oxygen, silicon is the second most abundant element, making up one fourth of the earth's surface. Phosphorus, boron and, sometimes, aluminum are the major chemical elements added to silicon to produce semiconductors.

If a finished silicon semiconductor is properly processed and conditioned, its electrical characteristics remain very stable. The effects of temperature change become far less critical compared with devices made of selenium or germanium. Although silicon itself is capable of operating from -176°C to +350°C, a typical operating temperature range listed on a silicon semiconductor data sheet is from -55°C to +125°C. Operation at temperatures beyond this range may cause shifts in device characteristics and will no longer allow it to perform properly within circuit design limits. Its operating stability, wide operating temperature range and abundance makes silicon the preferred base element used for semiconductor production.

# GALLIUM ARSENIDE, GALLIUM PHOSPHIDE AND GALLIUM ARSENIDE PHOSPHIDE

Combinations of gallium, arsenic and phosphorus produce gallium arsenide (GaAs), gallium phosphide (GaP) and gallium arsenide phosphide (GaAsP), materials used in the manufacture of other discrete semiconductors.

## MICROWAVE COMPONENTS

Gallium arsenide is characterized by a high molecular mobility that is superior to the molecular mobility of germanium. This material is used for diodes and transistors normally operating in the extremely high frequency (microwave) region.

## LIGHT-EMITTING DIODES (LEDs)

These devices have the capability of emitting a monochromatic, or single-frequency light when current of sufficient intensity is made to flow through them. The emitted light is either red, yellow, orange, green or nonvisible infrared, depending on the particular combination of materials used.

Light emitting diodes that emit visible light are replacing the traditional incandescent lights in indicator applications and are used in LED digital displays. The power consumed is about a thousand times less than that of an incandescent light used in a similar application, providing huge savings in power. In addition, the estimated life of a light emitting diode is approximately one million hours (114 years) compared with an average lifespan of two years for the incandescent light.

The nominal operating temperature range for this group of components ranges from -55°C to +100°C, limited only by the operating temperature capability of its normally used plastic package. If the LED chip is enclosed in an appropriate glass or metal package, it can operate in the temperature range of -55°C to +125°C.

A nonvisible infrared LED is used as part of an *opto-isolator* (opto-coupler) and as part of a *solid-state relay* (SSR). Opto-isolators and SSRs are actually hybrid circuits that consist of semiconductors and passive components enclosed in a single package, but are considered to be a single device.

These optoelectronic components and circuits are covered in PART II — OPTOELECTRONICS.

# THE SEMICONDUCTOR MECHANISM

The "semiconductor mechanism" is the action of a component made of a solid-state material that responds to the polarity of a voltage by changing its characteristics to make it work as a conductor or a nonconductor. Any component functioning in this manner is classified as a semiconductor.

The semiconductor mechanism is inherent in all semiconductors and is based on applying specific D.C. voltages to the device—*forward bias* and *reverse bias* voltages.

FORWARD BIASED (CONDUCTING) STATE
In Figure 3a, a semiconductor is shown in block form in series with a load resistor, $R_L$, connected to a D.C. voltage source, E.

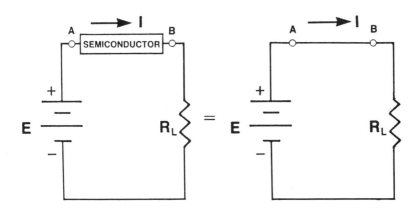

**Figure 3a**                    **Figure 3b**
The Semiconductor in its Conducting or Forward Biased Mode

With the polarity of voltage as shown and the semiconductor connected in the circuit in a way not yet specified, *the semiconductor takes on the characteristic of a conductor*.

Since a conductor is made of a material having essentially zero resistance, the circuit in Figure 3b is equivalent to the original circuit in Figure 3a and consists of the voltage, E, directly connected to the load resistor, $R_L$. Current flowing in the circuit is determined by the Ohm's Law relationship.

In a circuit, current, I, is equal to the supply voltage, E, divided by the load resistance, $R_L$.

$$I = \frac{E}{R_L}$$

In semiconductor terminology, the voltage applied between a semiconductor terminal and the circuit's reference point (ground) is referred to as the *bias voltage*. Because the device is in its conducting state and allows current to flow in the circuit, the semiconductor is said to be *forward biased*.

REVERSED BIAS (NONCONDUCTING) STATE
When the supply voltage is reversed, the voltage polarity on the semiconductor is also reversed. See Figure 4a.

The polarity reversal of the voltage connected to the semiconductor causes *the semiconductor to suddenly assume the characteristic of a nonconductor.*

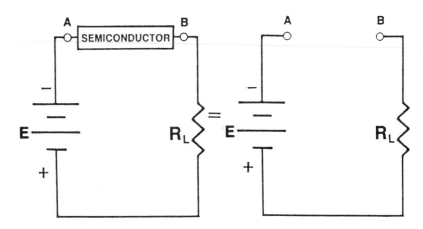

**Figure 4a**                    **Figure 4b**
The Semiconductor in its Nonconducting or Reversed Biased Mode

Since a nonconductor is a material having essentially infinite resistance, the semiconductor, acting as a nonconductor, causes the circuit to open, allowing no current to flow. The semiconductor is said to be in its nonconducting, or *reverse biased* mode. See Figure 4b.

# THE ELECTRONIC SWITCH

The inherent characteristics of the semiconductor that allow it to respond differently to positive or negative voltages, enable it to be used as an *electronic switch*.

This action can be compared to a single-pole, single-throw mechanical switch. It is either in its conducting state, with its contacts closed, or in its nonconducting state, with its contacts open. Unlike the action of a mechanical switch, a semiconductor responds to the polarity of a voltage, not to the movement of a mechanical lever, toggle, pushbutton or similar actuating mechanism. See Figures 5a and 5b.

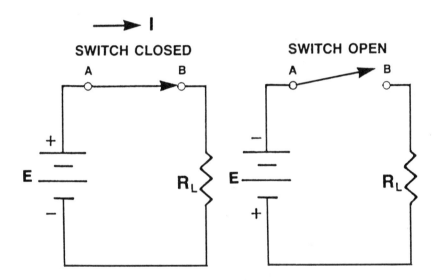

**Figure 5a**                    **Figure 5b**
Action of a Mechanical Switch

The response time of a mechanical switch does not compare favorably to that of a semiconductor. Typically, the response time of an ordinary semiconductor switch is about 100 nanoseconds, or 100 billionths of a second. Some semiconductor components, made specifically to function as fast switches, can respond in the order of one nanosecond and faster. These are the devices used in computer circuits and other high-speed digital switching applications.

# SUMMARY

The concept of creating a solid-state component that operates as a voltage polarity-sensitive switch is the basis of all semiconductor diodes. This concept is further expanded to develop all other semiconductor devices.

These other semiconductor components—thyristors, bipolar transistors and field effect transistors—have the capability of amplification. In some cases, they can be used as voltage-controlled resistors, in addition to functioning as switches.

Whatever its application, the basic semiconductor mechanism is at the heart of all semiconductor functions. Comprehension of this mechanism is the basis for understanding how all semiconductors work.

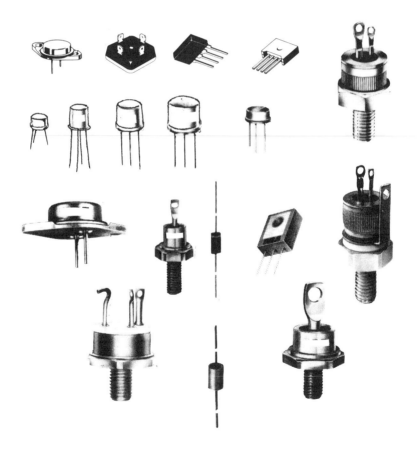

Discrete Semiconductors

# REINFORCEMENT EXERCISES

Answer TRUE or FALSE to each of the following statements:

1. A discrete semiconductor is a single, solid-state, active component in a finished package. It can operate as a conductor, non-conductor or as a variable resistor.

2. A discrete semiconductor is classified as an active device because it can differentiate between a positive or negative voltage polarity and respond differently to either polarity.

3. A component referred to as a semiconductor is designated that way because it only conducts current during half the time that voltage is applied to it.

4. Within the families of discrete semiconductors are diodes, thyristors, bipolar transistors, field effect transistors and integrated circuits.

5. Discrete semiconductors capable of dissipating 3 watts or more at 25°C are categorized as power devices. Discrete semiconductors with power ratings below 3 watts at 25°C are classified as small-signal devices.

6. The temperature range specified for discrete semiconductors intended for use in military and aerospace equipment is generally from -55°C to +125°C. Devices manufactured for use in commercial, industrial and consumer equipment normally are specified to operate within the range of 0°C to +70°C.

7. The chemical element germanium is the preferred material in the manufacture of discrete semiconductors because it is the most stable and most abundant of all available materials. It exhibits very little change in its characteristics when subjected to a high temperature environment.

8. Because of its extremely high molecular mobility, gallium arsenide is the material used in the production of discrete semiconductors to operate in the extremely high frequency (microwave) range.

9. A discrete semiconductor, in its forward biased state, will assume the characteristic of a conductor (zero resistance). In its reverse biased state, it will take on the characteristic of a nonconductor (essentially infinite resistance).

10. Any solid-state device that conforms to the action referred to as the "semiconductor mechanism" is considered to be a semiconductor. This term applies to the capability of the device to respond to a changing voltage polarity by changing from a conductor to a nonconductor and vice versa.

11. The inherent features of a semiconductor which allow it to change from a conductor to a nonconductor in response to a changing voltage polarity, make the device applicable for use as an electronic switch in a circuit.

12. When compared with a mechanical switch, the semiconductor is much slower in its response time.

Answers to reinforcement exercises on page 187.

CHAPTER
TWO

# DIODE PRODUCTION

WAFER PRODUCTION

WAFER PROCESSING

DIODE COMPLETION

# DIODE PRODUCTION

> A diode, in semiconductor form, is a two-element solid-state device that has the capability of functioning as a conductor and as a nonconductor. It is the least complex of all the components in the semiconductor family.

A diode can function as a:

- General purpose electronic switch
  - High speed computer switch
    - Converter of light to electricity
      - Rectifier that changes A.C. to pulsating D.C.
        - Voltage regulator and voltage reference
          - Circuit protector when used as a voltage clamp
            - Digital display element or as an indicator light

The diode can be more clearly understood if the technique of producing a rod and wafer of semiconductor-grade material and the succeeding steps needed to complete the final diode are examined.

Rod

Wafer

Chips

Beach                    Finished Diode

# STEP ONE - SILICON ROD PRODUCTION

## POLYCRYSTALLINE SILICON

Silicon is initially found in ordinary beach sand as silicon dioxide and is refined in high-temperature ovens into high-purity *poly-crystalline* silicon in the form of long, slender rods. The term "polycrystalline" describes its nonuniform lattice structure.

## SINGLE-CRYSTALLINE SILICON

The polycrystalline rod needs to be changed into the *single-crystalline* silicon rod required for the manufacture of semiconductor components. A single-crystalline structure is one in which the formation of the lattice is identically oriented throughout the crystal. The process of changing polycrystalline to single-crystalline silicon is accomplished by either of two techniques— the *float-zone* or the *Czochralski* method.

### FLOAT-ZONE METHOD
- A polycrystalline silicon rod is suspended in an oven having an oxygen-free atmosphere. The bottom of the rod is melted by a high-frequency induction coil to form a drop that hangs from the bottom of the rod.

- A seed crystal is then introduced into the silicon drop. With careful temperature control of the melted section, silicon is chemically grown on the seed in the form of a single-crystalline structure.

- As the R.F. induction coil is slowly raised along the length of the rod, the molten region of silicon progresses along the rod. This causes the silicon to dissolve and grow onto the seed to form a silicon ingot with a single-crystalline lattice structure.

### CZOCHRALSKI METHOD
- A bar of polycrystalline silicon is melted in the oxygen free atmosphere existing within a high-temperature oven.

- A slender single-crystalline bar, or seed, is introduced into the melt causing a portion of the melt to freeze onto the seed. As the process continues, the liquid silicon in the melt is formed into a cylindrical ingot of pure single-crystalline silicon.

## STEP TWO - WAFER PRODUCTION

The ingot, made by either method, is machined to form a rod of uniform diameter. It is then sawed into thin slices, or wafers, etched and lapped to remove any damage done by the sawing and polished to a mirror finish on one side. See Figure 6.

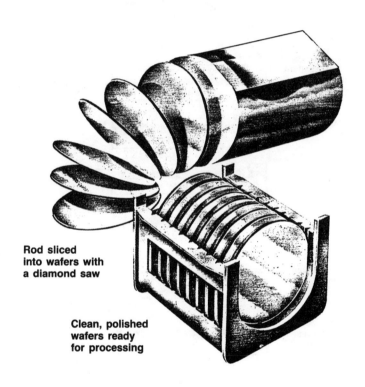

**Rod sliced into wafers with a diamond saw**

**Clean, polished wafers ready for processing**

Silicon Ingot Sawed into Wafers
**Figure 6**

Silicon ingot technology has progressed to the point where ingots of up to eight inches in diameter and up to 30 inches in length are being manufactured. Made by companies specializing in this type of silicon production, the wafers sliced from the ingots are sold to the semiconductor manufacturers for further processing into finished components.

# STEP THREE - WAFER PROCESSING

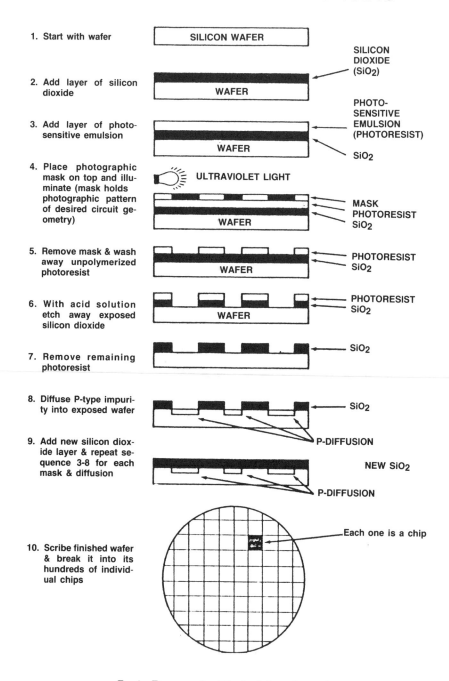

1. Start with wafer

SILICON WAFER

2. Add layer of silicon dioxide

WAFER

SILICON DIOXIDE (SiO2)

3. Add layer of photo-sensitive emulsion

WAFER

PHOTO-SENSITIVE EMULSION (PHOTORESIST)

SiO2

4. Place photographic mask on top and illuminate (mask holds photographic pattern of desired circuit geometry)

ULTRAVIOLET LIGHT

WAFER

MASK
PHOTORESIST
SiO2

5. Remove mask & wash away unpolymerized photoresist

WAFER

PHOTORESIST
SiO2

6. With acid solution etch away exposed silicon dioxide

WAFER

PHOTORESIST
SiO2

7. Remove remaining photoresist

SiO2

8. Diffuse P-type impurity into exposed wafer

SiO2

P-DIFFUSION

9. Add new silicon dioxide layer & repeat sequence 3-8 for each mask & diffusion

NEW SiO2

P-DIFFUSION

10. Scribe finished wafer & break it into its hundreds of individual chips

Each one is a chip

Basic Process in Diode Manufacturing
**Figure 7**

## EPITAXIAL LAYER

At the initial stage of wafer processing at the semiconductor manufacturer's facility, an optional processing step may be taken by depositing a thin layer of silicon on one surface of the wafer. This new layer, called *epitaxy* or the *epitaxial layer*, has the same crystalline structure as the original wafer. It is used to prepare the wafer to readily accept the *diffusion* of additional chemical elements later in the process.

The wafer acts as the host surface for the epitaxial layer and as the mechanical support for the finished diode. It is referred to as the *substrate*.

## PASSIVATION (GLASSIVATION)

The wafers are put in a high-temperature oven in the presence of oxygen. The combination of silicon and oxygen causes a layer of silicon dioxide to grow on the surface of the wafer. This is called the *passivation*, or *glassivation* layer and is used to isolate the diffusion steps from each other. See Figure 7.

## PHOTOMASKING

### COATING THE WAFER
In preparation for the photograph of the first mask, the passivated surface of the wafer is coated with *photoresist*, an emulsion that is sensitive to ultraviolet light.

### MAKING THE MASKS
Masks are flat glass plates used to create patterns on the wafer. They are used in the photographic process for each diode element. The use of a mask is comparable to the function of a template or stencil.

- Two large drawings are made, each representing the layout of each of the two elements of the diode. The layouts outline the areas for diffusion. Each drawing is reduced about 200 to 500 times through a photographic *step and repeat* process. The pattern of each drawing is repeated to produce a group of identical drawings on each mask.

- The photographic reproduction of the identical group of drawings on each mask is made to be the same size as the wafer. Each individual section is a reduced version of a large drawing containing the layout of each diode element.

Each mask contains several accurately spaced markings, called *registration marks*, that provide precise alignment of the masks on the wafer during the photographic process.

The number of individual reproductions on each mask depends on the size of the silicon wafer and the geometry of the completed diode. Small-signal diodes use about .010″ by .010″ in area. Power diodes require larger geometry using more silicon space, or more "real estate."

CREATING THE WAFER PATTERN
- The first mask is placed on the photoresist-coated surface of the wafer. The clear part of the masked pattern is exposed to ultraviolet light that *polymerizes* (fixes) the exposed photoresist areas on the wafer.

- The mask is removed and the exposed silicon dioxide areas are etched away with a sulphuric acid solution. The remaining photoresist and the remnants of the acid solution are washed away in preparation for the wafer to accept the first dopant into the exposed silicon areas through the process of doping or diffusion.

## DIFFUSION

Diffusion, or *doping*, is the process used to change the electrical characteristics of a wafer by having chemical elements, called *impurities* or *dopants*, penetrate the pure, single-crystalline wafer at high temperature.

Boron and phosphorus are the elements used as the dopants to produce the two sections of a silicon diode.

- The blank silicon wafer is placed in an atmosphere of boron in gaseous form inside a high-temperature oven (diffusion chamber). The boron is absorbed into the exposed silicon area producing positive ions in the silicon. This diffusion step causes that section in the wafer to change from pure silicon, having inherently infinite resistance, to a doped material that is highly conductive (essentially zero resistance).

Since this doped silicon region contains positive ions, it is referred to as the positive or *P-region*.

- A new layer of silicon dioxide (passivation layer) is grown over the first diffusion layer. The process is repeated with the second mask and with phosphorus as the second dopant.

- The effect of the phosphorus diffusion changes the other section of pure silicon material to a highly conductive area containing negative electrons. This negative-type material is referred to as the negative or *N-region* and lies adjacent to the previously diffused P-region. The line between the two adjacent regions is referred to as the *PN junction*. See Figure 8.

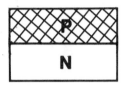

Single Chip PN junction
**Figure 8**

Ideally, all the chips of a finished wafer will have identical electrical characteristics. This can be achieved by precise alignment of the masks on the wafer, a carefully controlled diffusion process and the temperature maintained uniformly throughout the diffusion chamber.

## METALLIZATION

Attachment of the P and N regions on the chip to the external leads or terminals of the diode is facilitated by vaporizing a thin film of gold or aluminum (both highly conductive) to the designated areas, called bonding pads, on the diode elements. The *metallization* step completes the wafer fabrication process.

## WAFER SCRIBING AND SEPARATION

- The first step in wafer separation is the scribing of the wafer surface with a laser beam or a diamond-pointed scriber between the rows and columns of the diode patterns.

- The scribed wafer is then broken along the vertical and horizontal scribed lines into individual and identical diode chips, or dice.

# STEP FOUR - COMPLETION OF THE DIODE

## LEAD BONDING AND ENCAPSULATION

- One end of thin gold wires are attached to bonding pads on the chip by thermal compression or ultra-sonic techniques. The other ends are welded to external leads or terminals.

- The chip is then encapsulated in a hermetically-sealed glass envelope or other appropriate packages.

Typical Diode Packages

As a completed diode, the P-region is referred to as the *anode* and the N-region is called the *cathode*. The body of the package is branded with a stripe at the cathode end or with a properly polarized diode symbol. See Figure 9.

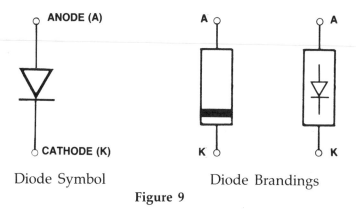

Diode Symbol          Diode Brandings

**Figure 9**

To verify compliance with data sheet specifications, diodes are subjected to all required tests. They are sorted and branded with a part number (See CHAPTER THREE for number designation.)

A diode intended for military or aerospace use will undergo a series of preconditioning and screening tests to meet the requirements of the appropriate military specifications. It is designated as a high reliability device and assigned a part number according to an appropriate military specification. See CHAPTER EIGHT — SEMICONDUCTOR RELIABILITY CONSIDERATIONS.

# REINFORCEMENT EXERCISES

Answer TRUE or FALSE to each of the following statements:

1. A semiconductor diode is a two-element, solid-state device that has the capability of functioning as a conductor and as a nonconductor. It is the least complex component in the semiconductor family.

2. Some applications of semiconductor diodes are for use as general purpose switches, high-speed computer switches, rectifiers, converters of light to electricity and voltage regulators.

3. For semiconductor-grade material, it is necessary to change the crystalline structure of silicon from polycrystalline to single-crystalline.

4. Most semiconductor companies in the United States produce their own silicon wafers to be used in the manufacture of their semiconductor products.

5. It is technically impossible to produce a silicon rod that is greater than three inches in diameter.

6. The chemical dopants that are added in the diffusion process to produce a semiconductor diode from silicon are boron and phosphorus.

7. With boron being diffused into silicon, negative or N-type regions are produced. The use of phosphorus as a dopant creates positive or P-type regions in the silicon wafer.

8. Thin layers of either gold or aluminum are used as bonding pads on the diode chip to facilitate attaching leads between the P and N regions of the chip and its external terminals.

9. After encapsulation in an appropriate package, the diode is tested, sorted and branded with a part number. In addition, a stripe is branded on one end of the diode package to indicate the anode terminal.

10. The P-region of the diode is designated as the anode and the N-region as the cathode.

Answers to the reinforcement exercises are on page 188.

CHAPTER
THREE

# DIODE CHARACTERISTICS AND SPECIFICATIONS

# DIODE CHARACTERISTICS
# AND SPECIFICATIONS

## DIODE ACTION

THE CONDUCTING OR FORWARD BIASED MODE
The action of a diode in its conducting or forward biased mode is
shown in the cross-sectional diagram of Figure 10a.

- A diode in series with a load resistor is connected to a D.C.
  voltage source. The positive terminal of the voltage source is
  connected to the P-region of the diode and the negative
  terminal of the voltage source is connected through the load
  resistor to the diode's N-region.

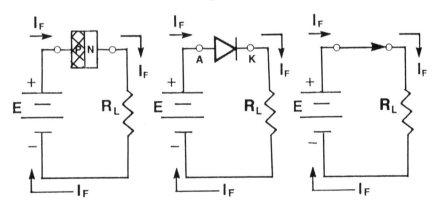

Cross-section Drawing    Schematic Diagram      Equivalent Circuit
   **Figure 10a**              **Figure 10b**                **Figure 10c**
          Conducting or Forward Biased Mode of a Diode

Under this condition of polarity (positive at the P-region and
negative at the N-region), the resistance of the PN junction is
essentially zero. The diode is in its forward biased or conducting mode, acting as a conductor.

In this state, the circuit is equivalent to a voltage source connected directly across a load resistor through the closed contacts of a
switch. The current, I, flowing in the circuit, is determined by the
value of the voltage, E, and the load resistor, $R_L$. See Figure 10c.

The cross-section of a diode, shown in Figure 10a, is replaced in Figure 10b with its graphic symbol.

• The P-region is shown as an arrowhead and is referred to as the *anode* (A). The direction of the arrowhead indicates the direction of current flow, called *forward current*, and is designated as $I_F$.

• The N-region, represented by a bar, is referred to as the *cathode* (K). By convention, the stripe branded on the body of the diode is at the cathode end.

THE NONCONDUCTING OR REVERSE BIASED MODE
• When the supply voltage is reversed, the negative terminal is connected to the P-region (anode) of the diode and the positive terminal is connected through the load resistor to the N-region (cathode). See Figures 11a and 11b.

• Polarity reversal causes positive ions at the PN junction to leave that area and move into the P-region.

• Simultaneously, negative electrons leave the PN junction and move into the N-region. The result of this action is the depletion of positive ions and negative electrons in the junction area, leaving a region of essentially pure silicon in the depleted section, called the *depletion region*. See the cross-section drawing in Figure 11a.

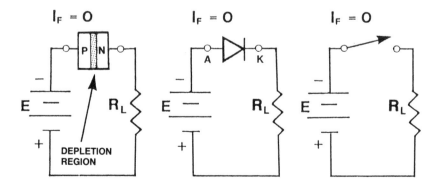

Cross-section Drawing   Schematic Diagram      Equivalent Circuit

**Figure 11a**          **Figure 11b**          **Figure 11c**

Nonconducting or Reverse Biased Mode of a Diode

With the creation of the depletion region, the resistance of the diode is essentially infinite, opening the circuit and allowing no current to flow. The diode is in its reverse biased or nonconducting mode, acting as a nonconductor.

In this state, the circuit is equivalent to a voltage source disconnected from its load through a set of open contacts of a switch. See Figure 11c.

# DIODE CHARACTERISTICS

## IN THE FORWARD BIASED REGION

If a semiconductor diode were an ideal device, the resistance of the PN junction would be zero and conduction would start when the smallest possible positive voltage is applied to the anode. See Fig. 12a.

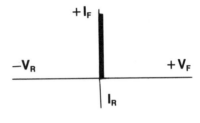

Ideal Diode
**Figure 12a**

Since a semiconductor diode is not an ideal device, a very small threshold voltage must be overcome before conduction can begin. See Figure 12b.

In a silicon diode, the threshold voltage is about 0.6 volts. In a germanium diode, the threshold voltage is about 0.15 volts.

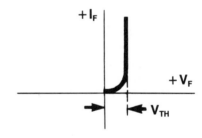

Threshold Voltage
**Figure 12b**

Since the semiconductor diode is not an ideal device, the resistance of the forward biased PN junction is not zero; it actually has a very small resistance value, $R_F$. When compared with the relatively large value of the load resistance, $R_L$, in the circuit of Figure 11, $R_F$ is negligible and is assumed to be zero. Typically, $R_F$ for small-signal diodes is about 2 ohms. For power diodes, $R_F$ is about .001 ohm.

When the forward voltage increases above the threshold
voltage, the diode appears as a
closed switch, since the load resistor, $R_L$, is considerably greater than $R_F$. The *forward current*,
$I_F$, in the circuit of Figure 10 is
determined by the value of the
supply voltage, E, and the load
resistor, $R_L$. See Figure 12c.

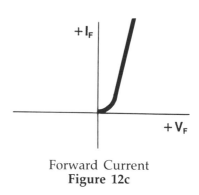

Forward Current
**Figure 12c**

When forward current flows through the diode, a small voltage
is created across the junction. This voltage is the product of the
forward current, $I_F$, and the small forward biased junction resistance, $R_F$. The supply voltage, designated as E, when used to
provide a forward bias voltage is called *forward voltage* and is
designated as $V_F$. $V_F = V_{TH} + (I_F \times R_F)$. The forward voltage is
typically below 1 volt. See Figure 12c.

IN THE REVERSE BIASED
REGION
If the semiconductor diode
were an ideal device, zero current would flow when the diode is reverse biased (negative
voltage is connected to the anode with respect to the cathode). See Figure 13a.

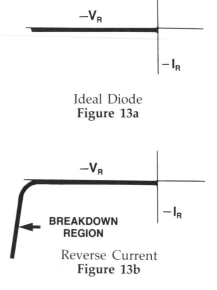

Ideal Diode
**Figure 13a**

Since the semiconductor diode
is not an ideal device, the resistance of the depletion region in
the reverse biased mode is not
quite infinity. Therefore, an extremely small *reverse current*, $I_R$,
flows in the circuit. See Figure
13b.

Reverse Current
**Figure 13b**

A typical value of reverse current is about 1 nanoampere (one
billionth of an ampere) for an ordinary silicon diode. For carefully processed silicon diodes, reverse currents of 1 picoampere,
and less, are attainable. The supply voltage, E, when used to
reverse bias the diode, is designated as the *reverse voltage*, $V_R$.

As the reverse voltage, $V_R$, increases, the depletion region narrows to the point where it no longer retains its original, almost infinite, resistance. The diode is now in its *reverse voltage breakdown* state and reverse current increases as the reverse voltage increases.

With a further increase in reverse voltage, the resistance of the depletion region further decreases, until its resistance decreases to essentially zero. In the diode's breakdown region, reverse current, $I_R$, is determined by the value of reverse voltage, E, and the load resistor, $R_L$. See Figure 13b.

Exceeding the power rating ($V_R \times I_R$) of the diode in its breakdown region can result in its immediate destruction. If the power rating in its breakdown region is not exceeded, the reverse voltage can be reduced to return the diode to its nonconducting state with no damaging effects.

# DIODE SPECIFICATIONS

Diodes are specified under the headings of *absolute maximum ratings* and *electrical characteristics.*

Absolute maximum ratings specify the maximum value of the voltage, current, power and temperature that the diode can safely handle. They define the electrical and thermal limitations of the device, however, diodes are rarely operated at the maximum values. Care must be taken to avoid having any combination of voltage, current and temperature exceed the maximum power rating. The absolute maximum ratings are specified on the data sheet to indicate the capabilities of the device and are often used as a measure of comparison with other diode types.

Electrical characteristics describe the diode under defined operating conditions and its significant characteristics are normally listed on the device data sheet. The applicable minimum and/or maximum values of the parameters are given as well as the typical values for characteristics under specifically defined conditions. These characteristics are often provided in the form of graphs, curves and charts.

## ABSOLUTE MAXIMUM RATINGS
### Specified at 25°C unless otherwise noted

CONTINUOUS FORWARD CURRENT ($I_{F\ cont}$)
This value is given in amperes or milliamperes. It depends on the size of the diode chip, the diameter of the leads from chip to the internal terminals and the diameter of the external leads. Package size can also influence this rating. Large chips in large packages can handle high currents up to 1500 amperes.

SURGE OR PEAK FORWARD CURRENT ($I_{F\ surge}$)
This current is based on a short duration pulse width (normally 1 microsecond) at 300 pulses per second. Different pulse widths, repetition rates and duty cycles can be used.

PEAK INVERSE (REVERSE) VOLTAGE (PIV or PRV)
   (See Fig. 12)
This rating, given in volts, specifies the maximum reverse voltage (negative voltage to the anode) that can safely be used and still have the diode maintain its nonconductive state. Silicon diodes are manufactured with ratings up to 3000 volts. To achieve very high PIV ratings, several diodes are connected in series as shown in Figure 14.

**ANODE**                                                                     **CATHODE**

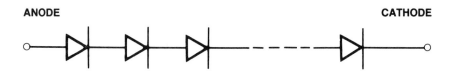

Achieving High Peak Inverse Voltage Ratings
**Figure 14**

D.C. POWER DISSIPATION ($P_D$) (Specified at 25°C)
The power dissipated by the diode is the product of the forward voltage ($V_F$) and the forward current ($I_F$). The maximum power capability of the diode depends on the size of the diode chip, the type and size of the package and its mounting features. Although the power rating is specified at 25°C, this temperature is rarely maintained. The diode power derating, given in watts or milliwatts, is merely a reference value and must be derated at elevated temperatures.

Power derating is the decrease in power dissipation rating in watts per degree C change in temperature from the rated power at 25°C. Derating information is provided either as a derating curve on a data sheet or as a derating factor.

As shown in the curve of Figure 15, if the diode is operated below 25°C, there is no increase in power capability.

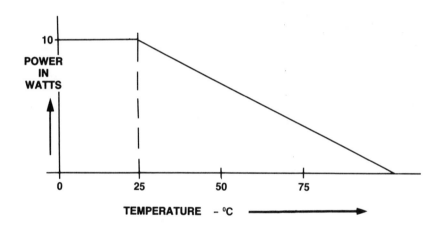

Power Derating Curve versus Temperature
**Figure 15**

Power diodes are available with ratings as high as several hundred watts at 25°C.

OPERATING AND STORAGE TEMPERATURE RANGE
For military and aerospace products: from -55°C to +125°C. There is no standard temperature range for diodes specified for commercial, industrial and consumer applications. Typically, the range is from 0°C to +85°C.

LEAD SOLDERING TEMPERATURE
This value depends on the type of device being used. Typically, it is a maximum of 270°C, if the soldering spot is at least 1/4″ from the seal between lead and package for no longer than 5 seconds.

## ELECTRICAL CHARACTERISTICS
Specified at 25°C unless otherwise noted

FORWARD VOLTAGE ($V_F$)
The forward voltage is the sum of the threshold voltage, $V_{TH}$, and the voltage across the forward resistance, $R_F$. It is specified at some value of forward current, $I_F$. Typically, it is less than 1 volt. $V_F = V_{TH} + (I_F \times R_F)$. The temperature coefficient in this region typically causes an increase in forward voltage of 2 millivolts per °C temperature increase.

REVERSE OR LEAKAGE CURRENT ($I_R$ or $I_L$)
Maximum and typical values of current flowing through the diode in its reverse biased, or nonconducting mode, are specified at the peak inverse voltage (PIV) value. Normally measured at a temperature of 25°C, $I_L$ is frequently specified at one or several other temperature values. A series of curves showing reverse currents versus reverse voltage at several temperature levels is often provided as part of a diode data sheet. See Figure 16. Typically, the positive temperature coefficient of the diode produces a doubling of leakage current for every 10°C increase in temperature.

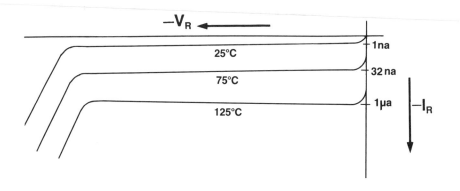

Reverse Current vs Reverse Voltage
**Figure 16**

The $I_R$ of a power diode can range from 1 to several hundred microamperes at 25°C, depending on the size of the chip. The $I_R$ of a small-signal silicon diode is typically 1 nanoampere at 25°C; a carefully processed small-signal silicon diode could be as low as 0.1 picoampere.

For either type, as the temperature increases, the value of $I_R$ increases, approximately doubling every 10 degrees C. The reverse resistance of the diode at any specified reverse voltage can be calculated as:

$$R_R = \frac{V_R}{I_R}$$

The ratio between the forward resistance, $R_F$, and the reverse resistance, $R_R$, is referred to as the *front-to-back-ratio* of the diode and indicates the quality of the device as a switch—its low resistance as a conductor compared with its high resistance as a nonconductor.

When characterized as a computer switch, the length of time it takes the diode to respond to a switching command to turn ON is called the *forward recovery time*. The time to turn the diode OFF is called the *reverse recovery time*.

### FORWARD RECOVERY TIME ($T_{fr}$)

At a defined reverse and forward voltage, this is the time it takes a diode in its nonconducting mode to revert to its stabilized conducting state after a forward-biasing digital pulse is applied. See Figures 17a and 17b.

### REVERSE RECOVERY TIME ($T_{rr}$)

At a defined value of forward current and switching voltage, this is the time it takes for a diode in its conducting mode to revert to its stabilized nonconducting state after a reverse-biasing digital pulse is applied. See Figures 18a and 18b.

For a diode to qualify as a computer switch, the reverse and forward recovery times must be 50 nanoseconds or less.

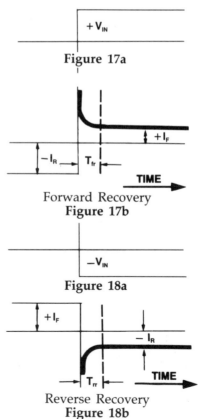

Figure 17a

Forward Recovery
Figure 17b

Figure 18a

Reverse Recovery
Figure 18b

## PHYSICAL CHARACTERISTICS

The information on package size, lead length and material, or fixed terminal size and material is provided in an outline drawing on a data sheet.

# ZENER DIODES

A zener diode is a specially processed silicon PN junction that maintains a predetermined constant voltage across its terminals despite changes in diode current. Because of this unique characteristic, it is used as a voltage regulator, a voltage reference or as a circuit protecting device.

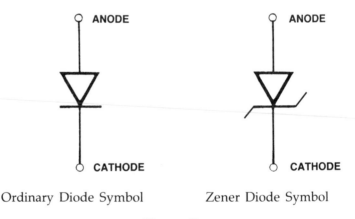

Ordinary Diode Symbol          Zener Diode Symbol

**Figure 19**

To differentiate a zener diode from an ordinary silicon diode, its schematic symbol is shown with "flags" at each end of the bar used to graphically designate the cathode symbol on a diode. See Figure 19.

Its constant voltage characteristic is achieved by applying a sufficiently high positive voltage to the cathode terminal of the diode with respect to its anode (reverse bias) to force the device to operate in its breakdown, or *avalanche* region. Operating in this region will maintain a constant D.C. voltage across the diode, regardless of any current variation through the diode. See Figure 20.

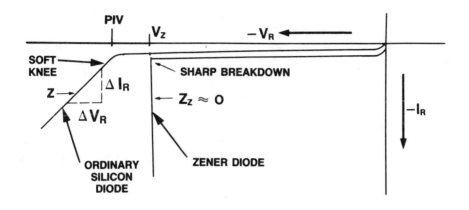

Zener Diode Characteristic
**Figure 20**

In Figure 20, there are two new symbols requiring definition:

- The Greek letter delta ($\Delta$) is used in electronics terminology to signify a changing quantity. The term $\Delta I_R$ means a change in reverse current and $\Delta V_R$ means a change in reverse voltage.

- The symbol ($\approx$) means "approximately". The term $Z_Z \approx 0$ means that $Z_Z$ is approximately equal to zero, or, it is close enough to zero to assume that it is equal to zero.

Unlike an ordinary silicon diode, the value of reverse voltage at which the zener diode will go into breakdown (avalanche) is predictable. This value is called the *zener voltage* and identifies the normal operating voltage of the device.

Although rarely operated in a forward biased state, if the zener diode is forward biased, it acts as an ordinary diode.

A typical breakdown curve of an ordinary silicon diode is illustrated in Figure 20. It is characterized by a "soft knee" appearance in moving from the high resistance reverse biased region into its breakdown region. The shape of the curve indicates an increasing voltage as reverse current increases. Its slope is referred to as a high *dynamic zener resistance*, defined as *the change in reverse voltage divided by the change in reverse current*. It is symbolized as $Z_Z$. In this case, the $Z_Z$ is a relatively high value.

The characteristic curve of a zener diode differs from that of an ordinary diode. When the reverse voltage is at the selected value of zener voltage, $V_Z$, the curve sharply changes over into its breakdown, or avalanche region. The shape of this curve is perpendicular (right-angle) to the horizontal axis, with the voltage remaining constant as reverse current increases.

As the zener current, $I_Z$, changes, there is essentially no change in zener voltage, making the slope of this curve (the dynamic zener resistance, $Z_Z$) essentially zero. This very desirable characteristic makes the zener diode most effective for its intended applications that include:

- Voltage regulation - Since the D.C. voltage across the zener diode in its breakdown (or avalanche) region is constant, any load connected in parallel with the zener diode will also be supplied with a constant or regulated voltage, regardless of changing input voltage or change in load conditions.

- Voltage reference - The use of a zener diode (or several zener diodes) as a voltage reference is characterized by its stable voltage regulating ability over a wide temperature range. Voltage reference diodes are useful in precision electronic equipment, such as ultra-stable power supplies, or in applications that requires the near-ultimate in voltage stability.

- As a voltage reference, the temperature coefficient (TC) of the zener crosses over from a negative value (at 3.2 volts to 4.6 volts) to a positive value (at 4.8 volts). The crossover point is about 4.7 volts or the zero TC point. Operating at this zener voltage (or multiples of 4.7 volts, if several of these diodes are connected in series) provides an extremely stable reference voltage, regardless of changing temperature. See Figure 21.

CATHODE                                              ANODE

4.7V        4.7V                          4.7V

Zero TC Diodes Connected in Series
**Figure 21**

- Voltage clamp - Zener diodes provide a means of limiting the voltage to a selected safe value at any desired point in a circuit. If connected into a circuit, the voltage at that point will not exceed the zener voltage, protecting that point from going above a limited value of voltage.

## ZENER DIODE SPECIFICATION RANGES

- Zener voltages: 3.2 volts to 220 volts in about 75 steps.

- Power capability: Zener voltage, $V_Z$, x Zener Current, $I_Z$. From 250 milliwatts to 250 watts at 25°C.

- Tolerances: 1%, 2%, 5%, 10% and 20%.

- Temperature Coefficient:
  NTC from 3.2 to 4.6 volts - from 0.04 to .001%/°C).
  At about 4.7 volts, the TC is zero.
  PTC from 4.8 to 220 volts - from .001 to 1.0%/°C.

- Operating and Storage Temperature Range:
  Military and space applications: -55°C to +125°C.
  Consumer, industrial and commercial applications:
  No standard range. Typically, from 0°C to 70°C.

# PART NUMBER DESIGNATION

The Joint Electron Device Engineering Council (JEDEC), an industry-sponsored organization, provides a vehicle for the standardization of diodes and other components. JEDEC is a major trade association, an arm of the Electronics Industry Association (EIA). The responsibilities of EIA include the formulation of standards of a technical nature, distribution of electronics marketing data and serves to interface between the electronics industry and various government agencies.

Each subcommittee set up by JEDEC is dedicated to a specific component category. Representatives from the manufacturers of these components and interested equipment manufacturers serve on these committees. Their major functions are to establish industry-compatible standards, component numbering systems, test methods, specifications uniformity, packaging standards and interchangeability criteria for components used in commercial, industrial and consumer applications.

These numbering systems and packaging standards are the basis for the numbering and packaging designations used in military type discrete semiconductors. See CHAPTER EIGHT — SEMICONDUCTOR RELIABILITY CONSIDERATIONS for information on military type part numbers.

JEDEC acts as a quasi-regulatory agency to monitor published component data sheet information and to recommend any revisions where conflicting JEDEC-registered numbers and specifications are concerned. JEDEC does not, however, control the actual specifications published by the component manufacturers nor monitor their adherence to JEDEC specifications.

JEDEC NUMBERING SYSTEM FOR DIODES
Example: 1N914

• The letter "N" signifies a semiconductor device.

• The prefix number "1" is used before the N to indicate a component with one PN junction (diode).

• The suffix number (914) is the JEDEC-assigned number for that diode and has no specific significance other than its sequentially assigned part number. Unless a user is very familiar with a part number and its specifications, characteristics of the diode can only be determined by studying the data sheet assigned to that part number.

In addition, JEDEC assigns a standard diode outline (DO) number to a registered diode to facilitate package interchangeability. For example, the diode outline number (DO#) of one manufacturer must be mechanically interchangeable with the diode of another manufacturer having the same DO number. The interchangeability of different diode packages can be determined by checking them against a standard DO number.

A JEDEC-assigned number does not imply any level of quality or reliability, nor does it imply that the JEDEC part number is any better or worse than a manufacturer's in-house part number. An in-house part number is often used rather than a JEDEC 1N part number so that proprietary control of the device can be maintained by either the manufacturer or customer.

# REINFORCEMENT EXERCISES

Answer TRUE or FALSE to each of the following statements:

1. A PN junction (semiconductor diode) is in its forward biased mode when the positive terminal of the supply voltage is connected to the P-region (anode) of the diode and the negative terminal of the supply voltage is connected through a load resistor to the N-region (cathode) of the diode.

2. The properties of a semiconductor diode are such that it will conduct current (essentially zero resistance at its junction) when reverse biased and will not conduct current (essentially infinite resistance at its junction) when forward biased.

3. The peak inverse voltage (PIV) rating of a diode refers to the maximum allowable reverse voltage that can be applied to the diode and still maintain operation of the diode within its nonconducting, or high resistance region.

4. The power capability of a diode at 25°C must be derated when the diode is operated at a higher temperature. If the diode is operated at a temperature below 25°C, power capability can be increased.

5. Subjecting the diode to a reverse voltage in excess of the peak inverse voltage will always put the diode into a failure mode and cause immediate destruction of the diode.

6. The front-to-back ratio of a diode refers to the ratio of its resistance in its forward conducting mode when compared with its resistance in its nonconducting state ($R_F/R_R$). The smaller the ratio, the more efficient the diode will be as a switch.

7. The reverse recovery time of a diode is specified as the time required for a diode in its conducting mode to revert to its stabilized nonconducting state after a reverse bias digital pulse is applied.

8. The forward recovery time of a diode is specified as the time required for a diode in its nonconducting mode to revert to its stabilized conducting state after a forward-biasing digital voltage step is applied.

9.  Special diodes are manufactured for high-speed switching use and are classified as computer diodes if their recovery time is 50 nanoseconds or less.

10. A zener diode is used for three major applications: as a voltage regulator, as a voltage reference source, and as a voltage clamp to protect any point in a circuit against excessive voltages.

11. A zener diode is generally operated in its breakdown or avalanche mode (reverse biased to its zener voltage value) and rarely in its forward biased state.

12. The higher the dynamic zener resistance, the more effectively the zener diode will function either as a voltage regulator, voltage reference or voltage clamp.

13. Zener diodes are available in many selections of voltage, voltage tolerance, current and power capability and in a wide range of operating temperatures.

14. Assigning a JEDEC "1N" number to a diode signifies that the device has only one element or one active terminal as part of its structure.

15. Assigning a JEDEC diode outline (DO) number to a diode package facilitates package interchangeability.

    Answers to the reinforcement exercises are on page 189.

CHAPTER
FOUR

# DIODE APPLICATIONS

GENERAL PURPOSE SWITCH

COMPUTER SWITCH

PHOTODIODE

VARACTOR

RECTIFIER

VOLTAGE REGULATION,
VOLTAGE REFERENCE AND
CIRCUIT PROTECTION

REGULATED D.C. POWER SUPPLY

REINFORCEMENT EXERCISES

# DIODE APPLICATIONS

## GENERAL PURPOSE SWITCH

A semiconductor diode can be used as a voltage-controlled switch to turn a circuit ON and OFF and has all the advantages inherent in semiconductor technology.

As a general purpose switch, the device:
- Is much faster than a mechanical switch.
- Is small, lightweight and uses no filaments.
- Is shock-proof and vibration-proof.
- Has essentially infinite longevity.

Depending on the characteristics and size of the chip and its package, a diode switch can handle large amounts of current (several hundred amperes) and very high voltages (up to several thousand volts).

A diode switch is in its ON state when a positive voltage is applied to its anode. See Figure 22a.

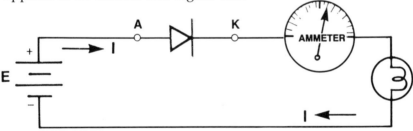

Switched ON Diode
**Figure 22a**

A diode switch is in its OFF state when a negative voltage is applied to its anode. See Figure 22b.

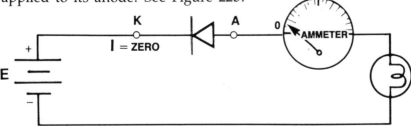

Switched OFF Diode
**Figure 22b**

## COMPUTER SWITCH

A computer circuit requires its switching elements to turn ON and OFF very quickly.

A small-signal diode can be used as the switching element of a computer if its switching time (reverse recovery and forward recovery) is 50 nanoseconds or less. See CHAPTER THREE – ELECTRICAL CHARACTERISTICS for a discussion of reverse and forward recovery times of a computer diode.

## PHOTODIODE

Generally, the glass envelope of a diode is coated with opaque paint to prevent light from striking the light-sensitive chip. The protective coating prevents the light from decreasing its very high reverse resistance characteristic.

If the diode is to be used as a photo-sensitive device, no opaque coating is applied to its clear glass envelope. When light strikes the diode chip, light current flows in the diode because of the photo-voltaic effect. Silicon and germanium are photo-voltaic materials and will act to convert light to electricity when exposed to a light source. See Figure 23.

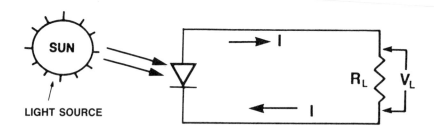

Photodiode Application
**Figure 23**

Because of the photo-voltaic effect, light or photo current will flow in the circuit with no need for a separate voltage supply. The voltage developed across the load (I x R) can be coupled to an amplifier to control the operation of a system. In this application, the photodiode can control the operation of a circuit by the presence or absence of light. Solar cells in calculators and solar power generators are two examples of products that utilize the photo-voltaic effect.

## VARACTOR (VARIABLE CAPACITOR)

When a diode is reverse biased, a depletion region is created (nonconducting section) between the P-region and N-region, resulting in the formation of a capacitor. The basic structure of a capacitor consists of a material with infinite resistance sandwiched between two conducting areas.

The measure of the storage capability of a capacitor is called its capacity. In this case, the capacity of the diode is very small (measured in picofarads) and will decrease as the reverse voltage increases.

As the reverse voltage varies around a fixed value, the *varactor* will act as a variable capacitor. Unlike an ordinary variable capacitor whose value is changed by the mechanical rotation of a shaft, the capacity of a varactor is controlled by a variation in voltage. See Figure 24.

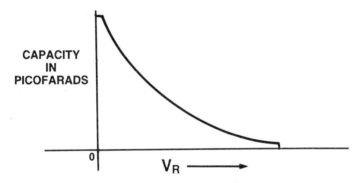

Varactor Diode Mechanism
**Figure 24**

The varactor diode can be used to control tuning in a radio receiver or transmitter as part of an *automatic frequency control* (AFC) circuit and operates as follows:

- A circuit can be monitored so that the value of the reverse voltage applied to a varactor will change when any unwanted temperature fluctuations and supply voltage variations occur.
- The desired direction of the change in reverse voltage will alter the value of the varactor's capacity, compensating for unwanted drifts in the circuit's parameters.
- If the varactor is part of a tuned, or resonant circuit, the proper setting of the circuit is maintained and automatically controlled, preventing any change in the tuned circuit.

## HALF-WAVE RECTIFICATION

In the circuit of Figure 25a, a diode in series with a load is connected to a source of A.C. voltage. The diode acts as a *half-wave rectifier*, a device that will change A.C. to half-wave pulsating D.C.

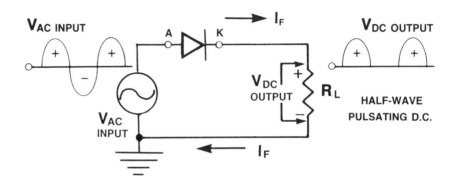

Half-wave Rectification (A.C. to Half-wave Pulsating D.C.)
**Figure 25a**

During the positive half-cycle interval of the applied A.C. voltage, $T_1$, a positive voltage is applied to the anode of the diode. During this time, the diode is forward biased and is in its conducting state.

Assuming the small voltage across the forward biased diode to be negligible, the equivalent circuit consists of a load, $R_L$, connected across a supply voltage, E. The load voltage, $V_L$, across $R_L$, is equal to the supply voltage, E. The current, $I_L$, flowing in the load is equal to the supply voltage, E, divided by the resistance of the load, $R_L$.

$$I_L = \frac{E}{R_L}$$

Throughout the negative half-cycle interval of the applied A.C. voltage, $T_2$, a negative voltage is applied to the anode of the diode. During this time, the diode is reverse biased and remains in its non-conducting state.

Assuming infinite resistance in the reverse biased diode, an open circuit exists with zero current flowing in the circuit and with no voltage across the load resistor.

The dual-polarity A.C.voltage at the source is changed to a single-polarity D.C. voltage, in this case, a continuing series of a single positive pulse for every cycle. Since only half of the A.C. is being used, this process is referred to as *half-wave rectification* with the resulting D.C. referred to as *half-wave pulsating D.C.*

In Figure 25a, the load voltage is positive with respect to the reference (ground). To obtain a negative D.C. voltage across the load resistor with respect to ground, the diode is simply reversed. See Figure 25b.

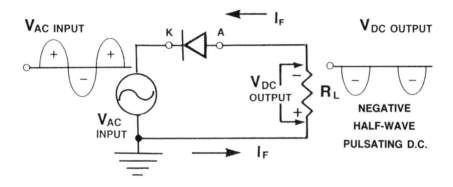

Half-wave Rectification with Negative Voltage Across the Load
**Figure 25b**

In this circuit, the diode will not conduct during the time interval, $T_1$, the duration of the positive half-cycle, but will conduct during $T_2$, the duration of the negative half-cycle.

To change half-wave pulsating D.C. to steady-state D.C. (the voltage source needed for electronic circuits), a capacitor filter circuit is used. The action of the filter is discussed in VOLUME ONE — CHAPTER TEN — CAPACITORS.

In many power supply applications, a half-wave rectifier is sufficient, with only a single diode used for rectification. Because only one-half of the applied A.C. voltage is used, very large filter capacitors are necessary to achieve an effective steady-state D.C. voltage. To reduce the need for large filter capacitors to achieve the required filtering action, *full-wave rectification* is preferred.

# FULL WAVE RECTIFICATION

### USING TWO DIODES AND A CENTER-TAPPED TRANSFORMER SECONDARY

In the full-wave rectifier circuit shown in Figure 26, the outer terminals of a power transformer secondary winding are connected to the anodes of diodes $D_1$ and $D_2$. The secondary center-tap is connected to one end of the load resistor, $R_L$. The cathodes of the diodes are connected to each other and then connected to one side of the load resistor. The reference is connected to the common connection between the load, $R_L$, and the transformer secondary center-tap (CT).

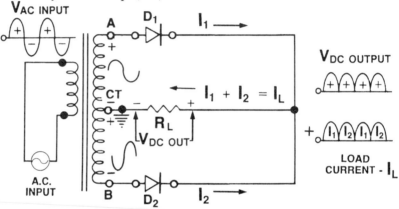

Full-wave Rectifier (With A Center-Tap Secondary)
**Figure 26**

With the diodes and load ignored and with A.C. applied to the primary winding of the transformer, two A.C. voltages exist across the secondary winding with respect to ground:
- the voltage between terminal "A" and the center-tap (CT).
- the voltage between terminal "B" and the center-tap (CT).

Each of these voltages is one-half the voltage across the entire secondary (from terminal "A" to terminal "B").

### SECONDARY VOLTAGE POLARITIES DURING THE POSITIVE HALF-CYCLE
- During this interval, the voltage at the top terminal of the secondary (terminal "A") is *positive* with respect to "CT".
- At the same time, the voltage at the bottom terminal of the secondary (terminal "B") is *negative* with respect to "CT".

As the A.C. voltage applied to the primary alternates, the polarity of the secondary voltages also alternates.

## SECONDARY VOLTAGE POLARITIES DURING THE NEGATIVE HALF-CYCLE
- During this interval, the voltage at terminal "A" becomes *negative* with respect to "CT".
- At the same time, the voltage at terminal "B" becomes *positive* with respect to "CT".

## FULL-WAVE RECTIFICATION MECHANISM
The polarity of these two voltage keeps alternating as the A.C. voltage of the power source alternates at the frequency of the power line (e.g. 60 Hz). When the diodes and load are connected to the transformer, as shown in Figure 26, the polarity of the voltages applied to the diodes changes as the A.C. changes.
- During the positive half-cycle, a positive voltage is applied to the anode of diode $D_1$, forward biasing $D_1$.
- At the same time, a negative voltage is applied to the anode of diode $D_2$, reverse biasing $D_2$.

As diode $D_1$ is forward biased, it acts as a conductor. At the same time, diode $D_2$ is reverse biased, making it act as a nonconductor. Only the upper half of the circuit is operating with the infinite resistance of diode $D_2$ effectively opening the lower portion of the circuit.

In the circuit of Figure 27a, current starts at terminal "A", flows through diode $D_1$, through the load, $R_L$, and back to the center-tap of the secondary to complete the circuit. The flow of current produces a positive voltage pulse across the load with respect to the reference ($I_1$ x $R_L$), equal to one-half of the total secondary voltage.

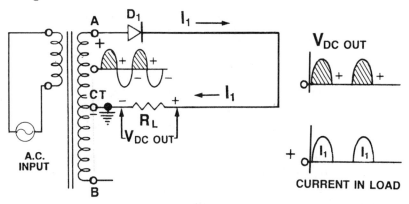

Conduction Through the Load During the Positive Half-cycle
**Figure 27a**

- During the negative half-cycle, a negative voltage is applied to the anode of diode $D_1$, reverse biasing $D_1$.

- At the same time, a positive voltage is applied to the anode of diode $D_2$, forward biasing $D_2$.

As diode $D_1$ is reverse biased (nonconducting) and diode $D_2$ is forward biased (conducting), only the lower half of the circuit is operating. The infinite resistance of diode, $D_1$, acts to effectively open the upper portion of the circuit.

In the circuit of Figure 27b, current starts at terminal "B", flows through diode $D_2$, through the load, returning to the secondary center-tap to complete the circuit.

The current produces a second positive voltage pulse across the load with respect to the reference ($I_2$ x $R_L$).

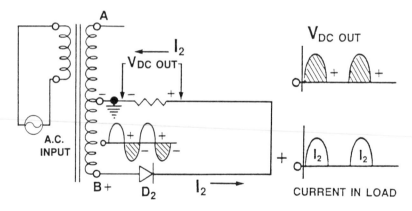

Conduction Through the Load During the Negative Half-cycle
**Figure 27b**

Through the action of two diodes and center-tapped transform-er, both halves of the A.C source voltage (positive and negative pulses) are changed (rectified) to two positive pulses for each cycle of the A.C. Since the full A.C. waveshape is used to produce rectified D.C., it is called *full-wave pulsating D.C.*

For each single A.C. cycle, this rectification process produces twice as many pulses during one time period of the cycle, effectively doubling the frequency of the applied A.C. voltage. To achieve the same degree of filtering needed for a half-wave rectifier circuit, full-wave rectification requires much less capaci-tance and is the preferred technique used for rectification.

# FULL-WAVE RECTIFICATION

## USING A BRIDGE RECTIFIER

If a transformer does not have a center-tap at the secondary winding, full-wave rectification can still be achieved with the use of a *bridge rectifier*. Four diodes are connected in a bridge configuration as shown in Figure 28. Two of the bridge terminals are connected to the transformer secondary terminals, another bridge terminal is connected to the top of the load resistor, $R_L$, and the last terminal of the bridge is connected to the bottom of the load resistor. This common connection is connected to the reference.

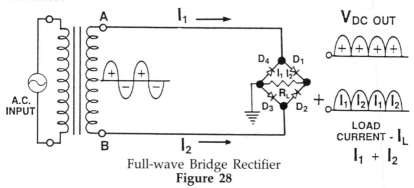

Full-wave Bridge Rectifier
**Figure 28**

Throughout the positive half-cycle of the A.C. power source, diodes $D_1$ and $D_3$ are forward biased (positive voltage on their anodes) and diodes $D_2$ and $D_4$ are reverse biased (negative voltage on their anodes).

During this period, the forward biased diodes, $D_1$ and $D_3$, are acting as conductors. Reverse biased diodes, $D_2$ and $D_4$, are acting as nonconductors, effectively opening the circuit in that section of the bridge rectifier. See Figure 29.

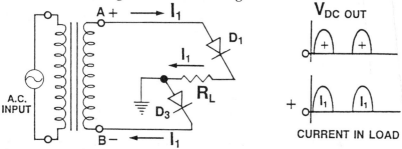

Conduction During Positive Half-cycle Using Bridge Rectifier
**Figure 29**

Tracing the current path from terminal "A" of the secondary
winding (positive during this interval of time), the current flows
through diode $D_1$ in the direction of the arrowhead, to the top of
$R_L$, through the load to diode $D_3$ in the direction of the arrowhead, to terminal "B" of the secondary winding to complete the
circuit. As the current flows through the load, it produces a
positive voltage pulse across the load with respect to the reference ($I_1 \times R_L$).

During the negative half-cycle of the A.C. power source, diodes
$D_2$ and $D_4$ are forward biased and diodes $D_1$ and $D_3$ are reverse
biased. During this period, diodes $D_2$ and $D_4$ are conducting,
while diodes $D_1$ and $D_3$ are in their nonconducting state, with
their infinite resistance effectively opening that portion of the
circuit. See Figure 30.

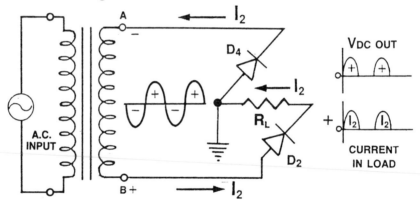

Conduction During Negative Half-cycle Using Bridge Rectifier
**Figure 30**

Tracing the current path from terminal "B" of the secondary
winding (positive during this interval of time), the current goes
through diode $D_2$ in the direction of the arrowhead, to the top of
$R_L$, through the load to diode $D_4$ in the direction of the arrowhead, to terminal "A" of the secondary winding to complete the
circuit.

As the current flows through the load, it produces another
positive voltage pulse across the load ($I_2 \times R_L$). For each A.C.
cycle at the voltage source, two positive pulses are produced
across $R_L$, changing A.C. to full-wave pulsating D.C.

Regardless of the method of rectification used, a negative pulsating D.C. voltage can be obtained across the load by reversing the
diode(s).

# ZENER DIODE APPLICATIONS

## VOLTAGE REGULATION

In Figure 31a, a 12-volt zener diode is chosen as the regulating device to provide a constant source of 12 volts steady-state D.C. The regulating circuit consists of input terminals connected to a resistor in series with a 12-volt zener diode. A source of unregulated steady-state D.C. voltage is applied across the input terminals, with its polarity as shown.

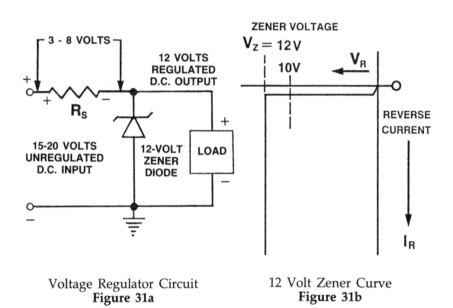

Voltage Regulator Circuit          12 Volt Zener Curve
**Figure 31a**                         **Figure 31b**

- If fewer than 12 volts are applied across the input terminals (plus polarity to the cathode of the zener diode), the zener diode is still in its nonconducting state and no current flows in the circuit. Under this condition, the full applied voltage is across the zener diode. See Figure 31b.

- If the supply voltage is increased to just below 12 volts, the diode is still in its nonconducting state and the applied voltage is still across the diode.

- If the supply voltage is increased to 13 volts, the zener diode snaps over to its breakdown region at the 12 volt point and current flows in the circuit. The voltage across the zener diode is now 12 volts.

In a series circuit, the sum of all the individual voltages in the circuit is equal to the supply voltage. If a supply voltage is 13 volts and the zener voltage is 12 volts, the voltage, $V_S$, across the series resistor, $R_S$, is 1 volt. See Figure 31a.

- If the supply is increased to 15 volts, the current in the zener diode would increase, however, the voltage across the zener diode would remain constant at 12 volts and the voltage across $R_S$ would increase to 3 volts.

- If the supply voltage is increased to 18 volts, there would be a further increase in zener current, with the voltage across the diode still remaining constant at 12 volts.

- If the supply voltage is increased to 20 volts, the voltage across the zener diode would remain constant at 12 volts and the voltage across $R_S$ would increase to 8 volts.

Series resistor, $R_S$, must be included in the circuit. It acts to absorb the voltage difference between the supply and the zener voltages. Without this resistor, current in the activated zener could not be limited and the diode would be destroyed.

In this circuit, if the supply voltage is set to vary from 15 to 20 volts, the voltage across the diode will always remain at 12 volts. Since a zener diode provides a constant voltage at a specified value (in this case, 12 volts), any load in parallel with the zener diode will have the same value of voltage applied. Regardless of variation in supply voltage or load resistance, the output voltage will always remain constant and voltage regulation will be achieved.

SUPPLY VOLTAGE RANGE
The power capability of the chosen zener diode must be large enough for the power rating of the diode not to be exceeded under *worst case conditions*. This term refers to the maximum power dissipated at the highest temperature of operation and indicates the upper limit of the supply voltage.

If the supply voltage decreases to below the value of the zener voltage (12 volts in the example given), the circuit will go out of regulation. The zener voltage value is the lower limit of the required supply voltage. As long as the supply voltage stays within the prescribed range of its upper and lower limits, regulation will always be maintained.

# VOLTAGE REFERENCE

To create a voltage reference equal to the value of a selected zener diode, a circuit consisting of a resistor in series with a zener diode is connected to a source of steady-state D.C. voltage that is greater than the zener voltage. See Figure 32.

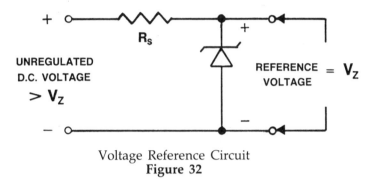

Voltage Reference Circuit
**Figure 32**

The voltage across the diode is the reference voltage used to compare with other voltages in a circuit. The mechanism of the zener diode used as a reference voltage is the same as the mechanism of the zener diode used as a voltage regulator. The major difference is that the zener used as a voltage reference has an extremely low temperature coefficient, producing very little or no change in voltage with changing temperature.

# CIRCUIT PROTECTION

Voltage surges, or transient voltage spikes, that inadvertently occur in a circuit or power line may be harmful to electronic equipment. Since these surges are generally short in duration, fast acting zener diodes are needed to provide a means of protecting circuit components against these excessive voltages.

PROTECTION AGAINST EXCESSIVE PLUS OR
MINUS D.C. VOLTAGES
To protect against excessive positive or negative voltages at any point in a circuit, the voltage at those points must be limited to a specified safe value. The zener diode selected depends on the value of voltage that must not be exceeded.

The selected zener diode is connected across the point(s) in the circuit requiring protection and polarized so that the voltage at its cathode is always positive with respect to its anode. See Figures 33a and 33b.

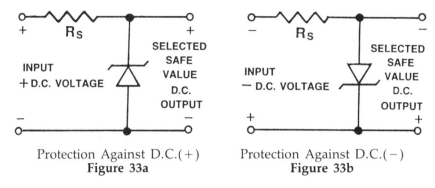

Protection Against D.C.(+)          Protection Against D.C.(−)
**Figure 33a**                      **Figure 33b**

The zener diode is connected in the circuit as a safeguard against any voltage above a specified safe value. When a voltage higher than the selected zener voltage is applied to the circuit, the zener diode will suddenly move into its constant voltage breakdown (or avalanche) region.

● This action will clamp the protected point in the circuit to the zener voltage, preventing any higher voltage spike or voltage surge from getting beyond that point.

● During normal circuit operation, with any reverse bias voltage on the zener diode kept below its breakdown voltage value, the diode will have no effect on the circuit.

## PROTECTION AGAINST EXCESSIVE PLUS AND MINUS D.C. AND A.C. VOLTS

The same principle of zener protection is used for a circuit subjected to random positive and negative D.C. voltage surges or transients and/or A.C. line voltage transients and surges.

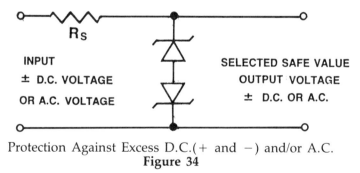

Protection Against Excess D.C.(+ and −) and/or A.C.
**Figure 34**

In this circuit, the protecting zener diodes are connected in series and in reverse polarity to each other. During the positive voltage transient, diode $D_1$ is clamped at its zener voltage value and diode $D_2$ is forward biased to its conducting state. During the negative voltage transient, the opposite actions occur.

## CONCEPTUAL DESIGN
## OF A REGULATED D.C. POWER SUPPLY

If unregulated A.C. is used as the prime source of power for electronic circuits and systems, a regulated D.C. power supply is required as the source of steady-state D.C. voltage. Whether the circuit supplying regulated D.C. voltage is external to the equipment or regulation is provided as part of the equipment, it is a major requirement of the electronic system.

In its basic form, a regulated D.C. power supply consists of four separate sections:

- Transformer
- Rectifier
- Filter capacitor section
- Voltage regulator

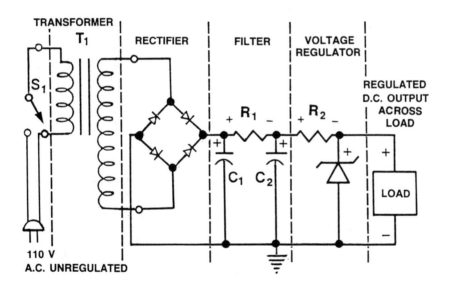

A Typical Regulated D.C. Power Supply
**Figure 35**

TRANSFORMER
This component changes the level of the A.C. line voltage connected to the primary winding of the transformer to the desired value of A.C. voltage at the output of the secondary winding(s). In addition, the use of a transformer provides electrical isolation between the lethal power line and the electronic circuit, providing protection to anyone who may come in contact with the circuit.

RECTIFIER

When used for either half-wave or full-wave rectification, the sole function of the rectifier diode(s) is to change the A.C. voltage coming from the secondary of the transformer to unregulated pulsating D.C. In the circuit shown in Figure 34, the diodes are connected as a full-wave bridge.

Since the A.C. is unregulated, the resultant pulsating D.C. is also unregulated and will vary in amplitude as the power line voltage varies.

FILTER CAPACITOR SECTION

The filter section is used to change pulsating D.C. to steady-state D.C. voltage required to power an electronic circuit. In the circuit shown in Figure 34, a two-section capacitor filter section is used with a resistor between the two capacitors. For better filtering efficiency, an inductor may be used to replace the resistor.

The steady-state D.C. voltage at the output of the filter is still unregulated and will vary in amplitude as the power line voltage varies.

VOLTAGE REGULATOR

This section of the power supply is identical to the voltage regulating circuit shown in Figure 31b. Its function is to provide a regulated (constant voltage)·D.C. output regardless of the varying D.C. voltage coming from the filter section. Some small amount of ripple that remains after filtering will also be reduced by the action of the voltage regulator.

Any changes in load conditions will not affect the voltage level of the regulated output of the zener diode circuit within the defined limits of the design.

The circuit shown in Figure 35 illustrates a relatively simple regulated D.C. power supply. In VOLUME THREE — PART ONE: INTEGRATED CIRCUITS, additional and more sophisticated voltage regulator circuits are examined.

# REINFORCEMENT EXERCISES

Answer TRUE or FALSE to each of the following statements:

1. When used in a circuit as a general-purpose switch, a semiconductor diode offers the features of fast switching in addition to voltage-controlled operation capability. It is small, lightweight, shock and vibration-proof and has infinite longevity.

2. A diode switch is in its ON state when a positive voltage is applied to the anode terminal and is OFF when no voltage or a negative voltage is applied to the anode terminal.

3. The response times of a computer diode, called the forward and reverse recovery time, are defined as the time required to revert from the diode's nonconducting state to its conducting state and vice versa.

4. A photodiode will exhibit a photo-voltaic effect when an external source of light impinges on the light-sensitive surface of a chip inside the package. This phenomenon causes light current to flow in a circuit in which the photodiode is connected.

5. To produce current in a photodiode, a source of D.C. voltage must be used in conjunction with the photo diode circuit.

6. Since a diode forms a high-resistance depletion region between the P and N regions when it is reverse biased, it can be used as a small variable capacitor under this condition. In this application, it is referred to as a varactor and will increase in capacitance as the reverse voltage across its terminals increases.

7. One of the applications of a diode, or a group of connected diodes, is rectification. This is the process of changing an A.C. voltage to a pulsating D.C. voltage. In this use, the single diode or properly interconnected multiple-diodes are referred to as rectifiers.

8. Full-wave rectification is a more efficient technique for changing A.C. to D.C., since both halves of the A.C. wave are used to provide pulsating D.C. voltage at twice the A.C. power line frequency.

9. In using a bridge configuration as a full-wave rectifier, it is always necessary to use a transformer with a center-tapped secondary winding connected at the input to the bridge rectifier.

10. In selecting a zener diode, the minimum power rating required is determined by calculating the product of the zener voltage and the highest value of zener current (power dissipation) at the highest temperature of operation.

11. A zener diode can function as an effective voltage clamp at a specific point in a circuit. It will act to protect that part of the circuit from any voltage transient, spike or surge that is impressed on the protected part of the circuit.

12. A single zener diode, or two properly connected zener diodes, can be used to protect against excessive positive or negative D.C. voltages and/or an excessive A.C. voltage.

13. In its basic form, a regulated D.C. power supply consists of a transformer, a rectifier, a filter section and a voltage regulator.

Answers to the reinforcement exercises are on page 191.

CHAPTER
FIVE

# THYRISTORS

# THYRISTORS

A thyristor is a 3-terminal semiconductor switching device with separate input (control) and output (load) circuits. Relatively low control current causes the output section of the thyristor to be turned ON, allowing high current to flow in the load. Once the device is turned ON, the input section no longer has control of the device. Turn-off is controlled only by the output circuit supply voltage.

In comparison with the thyristor, the 2-terminal diode has no separate control circuit—its control and load current is the same current.

The two major components in the thyristor family are:
- The *silicon controlled rectifier* (SCR)
- The *bi-directional triode A.C. switch* (TRIAC)

Thyristors are generally called *A.C. switches* and are used in a variety of power applications.

## SILICON CONTROLLED RECTIFIER (SCR)

A silicon controlled rectifier, or SCR, is a uni-directional, 4-layer (P-N-P-N) switching device. See Figures 36a and 36b.

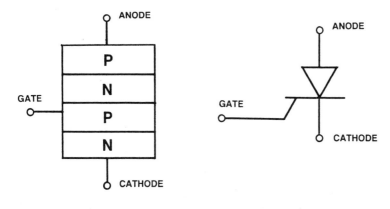

SCR Cross-section
**Figure 36a**

SCR Graphic Symbol
**Figure 36b**

The three terminals of the SCR are designated as:
- Anode - the output terminal
- Cathode - the reference terminal, or ground
- Gate - the input, or control terminal

In the circuit shown in Figure 37a, the D.C. supply voltage is designated as $V_{AA}$. In semiconductor terminology, any D.C. supply voltage is designated with *double subscript notation*. The double subscript letter is assigned by using the first letter of the terminal to which voltage is being provided with respect to the reference.

The letter, V, has two identical subscripts, AA, the doubled first letter of the anode. This is the terminal connected through the load to the supply voltage.

## SCR TURN-ON MECHANISM WITH D.C. SUPPLY VOLTAGE

As shown in Figure 37a, the gate is open, or unconnected.
- The positive terminal of the D.C. supply voltage is connected through the load, $R_L$, to the anode of the SCR; the negative terminal of the supply is connected to the cathode, which is connected to the circuit reference, or ground.
- Although a positive voltage is connected to the anode, with the gate open, no current will flow in the output circuit.

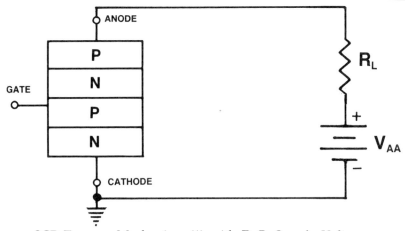

SCR Turn-on Mechanism (1) with D.C. Supply Voltage
**Figure 37a**

If a single PN junction existed, it would be forward biased and act as a conductor, however, the SCR is a 4-layer structure and only the top and bottom PN junctions are forward biased.

Since the top N-region is more positive than the lower adjacent P-region, the inner NP junction is reverse biased and there is infinite resistance between them.

To turn the SCR switch ON and complete the load circuit, it is necessary to supply current to the gate section.

- In the circuit of Figure 37b, an additional small D.C. voltage, $V_{GG}$, is applied through a large external resistor, $R_S$, to limit gate current.

- Once the SCR is turned ON, the gate no longer has control over the device.

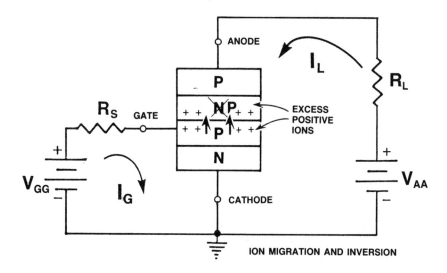

SCR Turn-on Mechanism (2) with D.C. Supply Voltage
**Figure 37b**

- The gate supply voltage, $V_{GG}$, is used to apply a forward bias at the PN junction (gate-to-cathode). The forward biased junction, now acting as a conductor, closes the gate-to-cathode circuit, allowing gate current to flow.

Compared with the load current (measured in amperes), gate current is relatively very low (microamperes to milliamperes). Gate current, $I_G$, is determined by gate voltage, $V_{GG}$, and the series resistor, $R_S$:

$$I_G = \frac{V_{GG}}{R_S}$$

- The flow of gate current forms excess positive ions in the gate terminal's P-region. These positive ions are attracted and migrate to the upper adjacent N-region.

This process is called *ion migration* and *inversion* and causes the original N-region to change to P-type material.

The inversion creates a new structure consisting of a single large PN junction.

- With positive anode voltage, the new PN junction is in a forward biased state, causing the output section of the SCR (anode-to-cathode) to act as a conductor. See Figure 38.

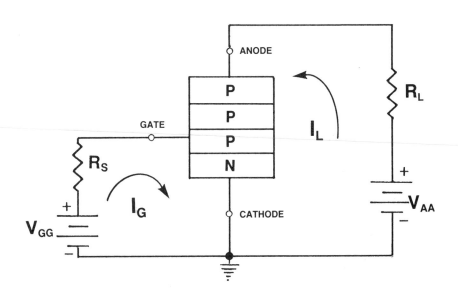

SCR Turn-on Mechanism (3) with D.C. Supply Voltage
**Figure 38**

- With the anode-to-cathode path closed, the output circuit is completed, allowing load current to flow.

Since the forward resistance of the SCR is essentially zero, the load current, $I_L$, is determined almost entirely by the supply voltage, $V_{AA}$, and the load resistor, $R_L$, and not by the SCR.

## SUMMARY OF THE SCR TURN-ON MECHANISM WITH D.C. SUPPLY VOLTAGE

| Anode Voltage | Gate (Control) Current | SCR Output Circuit |
|:---:|:---:|:---:|
| **Positive** | **Present** | **ON** |
| Zero | Present | OFF |
| Zero | Zero | OFF |
| Negative | Present | OFF |
| Negative | Zero | OFF |
| Positive | Zero | OFF |

The *simultaneous* conditions required for turn-on are:

- The anode must have a positive polarity voltage with respect to the cathode (reference).

- Gate (control) current must be present in the input circuit.

With zero, or negative anode voltage, the device will not turn ON, even if gate current is present. With gate current at zero, the device will not turn ON, regardless of the state of the anode voltage.

Once the SCR is turned on, it will stay in the ON state as long as the anode voltage remains positive, even if the gate current is zero. This is called *latch-up*, which means that once the device is turned ON, it will remain in the ON state with the gate no longer in control.

## SCR TURN-OFF MECHANISM WITH D.C. SUPPLY VOLTAGE

To turn the SCR OFF, the anode-to-cathode voltage must be reduced to zero or negative.

To prevent the SCR from turning ON again (once turned OFF):
- The anode-to-cathode voltage must be kept at zero or negative at any time that gate current is present, or,

- The gate current must be reduced to zero (and kept at zero) whenever the anode is positive with respect to the cathode.

## TURN-OFF TECHNIQUES

### USING A FIXED D.C. SUPPLY VOLTAGE (Figure 39a)

- If the SCR were turned ON with an input step voltage while there was positive D.C. voltage between anode and cathode, opening an external switch that is connected in series with the load would open the load circuit and turn the device OFF. As long as this switch is open, the SCR will stay OFF.

- If the gate current is removed, and stays removed, the switch in the load circuit can then be closed and the SCR will still remain OFF.

The switch must be capable of handling the normally high SCR anode current.

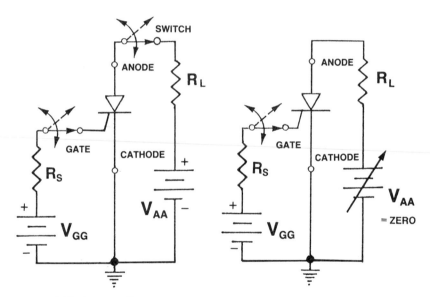

SCR Turn-off Mechanism with D.C. Supply Voltage

**Figure 39a**          **Figure 39b**

### USING A VARIABLE D.C. SUPPLY VOLTAGE - (Figure 39b)
The SCR will turn OFF when:
- The variable D.C. supply voltage, $V_{AA}$, is reduced to zero, reducing the anode-to-cathode voltage to zero.

- If the gate current is removed by opening the switch, the anode voltage may be raised, but the SCR will remain OFF.

## SCR MECHANISM WITH A.C. AS THE SUPPLY VOLTAGE

If the D.C. supply voltage is replaced with an A.C. supply, the SCR will turn ON if:
- The anode-to-cathode voltage is positive—this will occur during the positive half-cycle of the applied A.C. supply voltage (see Figure 40), and if:

- The input voltage is applied to the gate circuit allowing gate current to flow.

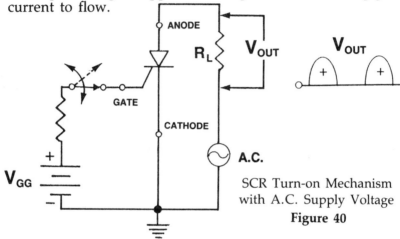

SCR Turn-on Mechanism with A.C. Supply Voltage

**Figure 40**

The SCR will turn OFF:
- When the A.C. voltage goes to zero or negative, regardless of the condition of the gate circuit.

The SCR will turn ON again:
- If the gate current is still present and the A.C. supply voltage becomes positive again.

The SCR will turn OFF and stay OFF when:
- The gate current is removed by opening the contacts of the SPST switch in the gate circuit, and when:

- The A.C. supply voltage reaches its zero value. The SCR will remain OFF regardless of the state of the anode voltage, positive or negative, as long as there is no gate current.

This characteristic of the SCR indicates that it is a *zero-crossing* turn-off device and will always turn OFF when the A.C. supply voltage moves from a positive voltage value to zero.

## SUMMARY OF THE SCR SWITCHING ACTION WITH A.C. SUPPLY VOLTAGE

- The SCR switch (anode-to-cathode) is turned ON when the anode voltage is positive and gate (control) current is present.

- The SCR switch is turned OFF when the A.C. supply voltage completes its positive half-cycle at the zero-crossing point remaining OFF during its negative half-cycle, even if gate current is still present.

- If the gate current is removed, the SCR will turn OFF and stay OFF when the A.C. supply voltage completes its positive half-cycle and crosses through zero.

- Output circuit conduction is initiated at the moment that gate current is applied during the positive anode voltage half-cycle. Conduction will continue until the anode voltage reduces to zero. This period is called the *conduction angle*. See Figure 41.

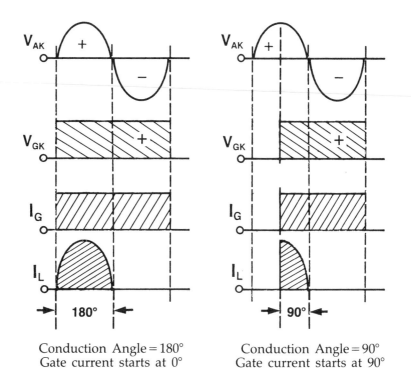

Conduction Angle = 180°          Conduction Angle = 90°
Gate current starts at 0°        Gate current starts at 90°

SCR Conduction Angle

**Figure 41**

# SCR SPECIFICATIONS

Data sheet specifications are classified under the headings of *absolute maximum ratings* and *electrical characteristics*.

Absolute maximum ratings are the maximum values assigned by the manufacturer, which, if exceeded, may result in permanent damage to the device. These ratings are used as limiting values and are not the actual operating conditions of the SCR.

Electrical characteristics are measurable properties of the device inherent in its design and are listed at specific temperatures. They describe the features of the device under conditions that approximate actual operation.

## ABSOLUTE MAXIMUM RATINGS

REPETITIVE PEAK REVERSE (INVERSE) VOLTAGE - PRV or PIV
Maximum allowable instantaneous repetitive reverse (negative) voltage that may be applied to the anode with the gate open. This rating is used when the supply voltage is A.C. Typical voltage ratings range from 25 to 2500 volts.

NON-REPETITIVE (TRANSIENT) PEAK REVERSE VOLTAGE - $PRV_{TRANSIENT}$
Maximum allowable instantaneous reverse voltage that may be applied to the anode with the gate open. The duration of the transient should not exceed 5 to 10 millisecs. Depending on the SCR, ratings range from 25 to 2600 volts.

CONTINUOUS REVERSE D.C.(BLOCKING) VOLTAGE - $V_{RDC}$
Maximum allowable reverse voltage that may be applied to the anode on a continuous basis (steady-state D.C.). Depending on the SCR, ratings may range from 40 to 2500 volts.

PEAK POSITIVE ANODE VOLTAGE - PFV
Maximum allowable forward voltage that may be applied to the anode with no gate current flowing. This rating defines the peak supply voltage, either A.C. or D.C. Depending on the SCR, typical values range from 500 to 1000 volts.

AVERAGE FORWARD CURRENT - $I_F$
Maximum continuous current that may be allowed to flow from anode to cathode under specified conditions of frequency, temperature and voltage. Standard devices are available with ratings from 1 ampere to 1000 amperes.

PEAK ONE-CYCLE SURGE CURRENT - $I_{SURGE}$
Maximum allowable non-recurrent peak forward current of a
single cycle (8.3 milliseconds in duration) in a 60 Hertz resistive
load circuit. Standard devices are available with ratings as high
as 3500 amperes.

PEAK GATE POWER DISSIPATION - $P_{GM}$
Maximum allowable power (gate voltage times gate current) the
gate circuit is capable of dissipating at 25°C. Depending on the
SCR type, these ratings can range from 0.1 watts to 400 watts for
a specified time duration in microseconds.

AVERAGE GATE POWER DISSIPATION - $P_{G(AV)}$
Maximum allowable power the gate circuit is capable of dissipating under continuous operation at 25°C. These ratings range
from 0.01 watts to 2 watts.

PEAK POSITIVE GATE CURRENT - $I_{GM}$
Maximum allowable gate current under continuous operation.
These values can range from 0.1 amperes to several amperes.

PEAK GATE VOLTAGE - $V_{GM}$
Maximum allowable gate voltage between gate and cathode
terminals. The values can range from 5 to 20 volts.

TOTAL POWER CAPABILITY (AT 25°C) - $P_T$
This rating specifies the maximum power that may be safely
sustained by the device when its case temperature is kept at 25°C
or lower. Values range from 450 milliwatts for low power devices
to over 1000 watts for extremely high power devices.

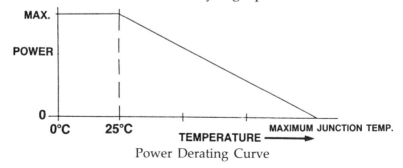

Power Derating Curve

OPERATING AND STORAGE TEMPERATURE RANGE
For military and aerospace applications: -55°C to +125°C.
For commercial, industrial and consumer applications, there are
no standard limits. The temperature range is determined by the
manufacturer and is typically -10°C to +100°C.

## ELECTRICAL CHARACTERISTICS

These parameters are listed at a specified temperature as minimum or maximum values, or both, and as typical values.

GATE TURN-ON (TRIGGER) CURRENT - $I_{GT}$
The value of gate current required to turn on the SCR at a specified value of anode to cathode voltage. For low-power devices, this is about 2 to 10 microamperes. For high-power devices, $I_{GT}$ is typically about 100 to 150 milliamperes.

GATE TURN-ON (TRIGGER) VOLTAGE - $V_{GT}$
The value of gate to cathode voltage needed to turn on the SCR at a specified value of anode to cathode voltage. Typically, these values range from 5 to 20 volts.

FORWARD VOLTAGE - $V_F$
The resistance between anode and cathode during the ON state is essentially zero, however, the small resistance that does exist, combined with the anode current, produces a voltage between these elements. This value is listed at a specified load current, typically ranging between 1.5 to 2.0 volts, similar to the voltage of a forward biased diode.

REVERSE (LEAKAGE) CURRENT - $I_R$
The value of anode current with reverse anode voltage and an open gate at a specified operating temperature. This is similar to the leakage current of a reverse biased diode. Typically, its value is about 2 microamperes for a low power SCR and 10 milliamperes for a very high power SCR at 25°C.

TURN-ON TIME - $T_{ON}$
This parameter consists of two time factors:
- Delay Time ($T_D$) - the interval between the initiation of the gate current and the increase in anode (load) current to 10% of its final value while switching from OFF to ON.
- Rise Time ($T_R$) - the interval between the end of the delay time to the point where the anode current reaches 90% of its final value (full conduction).

The sum of these two intervals is the turn-on time. It is measured at specified conditions of load current, gate current and temperature. For low-power SCRs, a value for $T_{ON}$ is typically 1.5 microseconds. For very high-power devices, this value is typically about 6 microseconds.

## TURN-OFF TIME - $T_{OFF}$

The interval between the time the anode voltage is reduced to zero and the device has actually returned to its OFF state. For low power SCRs, $T_{OFF}$ is typically about 40 microseconds. For very high power SCRs, $T_{OFF}$ is about 300 microseconds.

# ADDITIONAL DATA SHEET INFORMATION

## THERMAL RESISTANCE (JUNCTION TO CASE) $0_{J-C}$

The rise in temperature in °C between the anode and cathode (across all three junctions of the device) and the case, per unit power dissipation, $P_D$, in watts.

The major source of power dissipation is the product of the anode-to-cathode voltage during conduction and anode current.
- For low power devices, this main source of power is approximately equal to total power dissipation.
- For high power SCRs, gate dissipation must be included in the total power dissipation.

The heat developed by power dissipation flows from the combined three SCR junctions to the case. Junction temperature will rise above case temperature in direct proportion to heat generated.

$$\text{Thermal Resistance} = \frac{\text{Junction Temperature - Case Temperature}}{\text{Power Dissipation}}$$

Information on thermal resistance is necessary to calculate heat flow through the SCR and, in turn, determine the heat sinking and cooling equipment required for safe operation.

## CASE OUTLINE

The drawing of the package outline appears on the data sheet and includes information on package dimensions, mounting and torque details (sometimes required). To identify the SCR terminals, the package base configuration is also included.

Most SCR data sheets will contain curves of the operating characteristics under varying parameters of temperature, voltage, current and different conduction angles.

# SCR APPLICATIONS

Silicon controlled rectifiers are used in a variety of power control applications. These include:

• Lighting and heating
• Circuit protection against transient voltages
• Regulated power supplies
• Solenoid and power relay driver

## LIGHTING AND HEATING CONTROL

The circuit shown in Figure 42 is a basic SCR control circuit used for power control in lighting and heating systems.

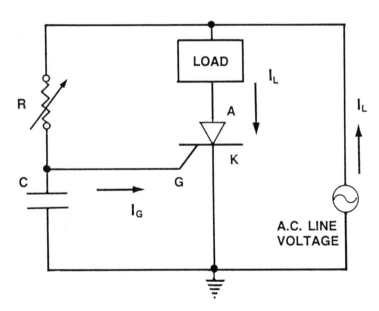

SCR Control Circuit for Lighting and Heat Control
**Figure 42**

The output section of this circuit is identical to the output circuit of Figure 40. In that circuit, a separate supply voltage is used to provide gate current to turn the SCR ON.

In the circuit of Figure 42, a single common power supply is used for both anode-to-cathode and gate-to-cathode voltages. Gate voltage is obtained by connecting a variable resistor in series with a capacitor across the A.C. supply voltage.

The voltage developed across the capacitor provides the forward bias for the PN junction between gate and cathode, with the variable resistor in the RC network used to vary the time for the A.C voltage to be applied to the capacitor. This time is determined by the value of C and the setting of the variable resistor, R. Forward bias of the gate-to-cathode circuit allows gate current to flow.

- With R adjusted to zero resistance, the A.C supply voltage will be in phase with the voltage applied to the capacitor. Since both the anode-to-cathode voltage and the gate-to-cathode voltage (capacitor voltage) start at the same point, the maximum conduction angle (180°) is achieved for the anode or load current. Under this condition, the highest average load current will be achieved.

- With R adjusted to a value greater than zero, the A.C. voltage will be applied to the capacitor at some time later than the zero point of the A.C. supply voltage cycle. When the value of the resistor, R, is increased, the load current conduction angle will change from 180° (R = 0) to 0° (R = ∞).

Although the value of R can be theoretically adjusted to infinity (∞), the largest practical value of R is some finite value that is considerably below infinity, providing a conduction angle somewhat greater than zero. At the maximum setting of R, the average load current will be at its lowest.


## LIGHTING CONTROL CIRCUIT MECHANISM

With an incandescent light (or a bank of lights) used as the load in the circuit of Figure 42, the value of the variable resistor, R, can be adjusted to provide either maximum light intensity or a dimmed lighting condition.

CASE ONE (with R = zero):
- With the variable resistor, R, set to zero, the voltage across the capacitor, C, is connected directly across the A.C. supply voltage. Under this condition, the capacitor voltage will be in phase with the A.C. supply voltage.

- As the A.C. supply voltage increases from zero into the positive half-cycle region, voltage is developed across the capacitor to provide the forward bias for the gate-to-cathode PN junction, allowing gate current to flow. See Figure 43a.

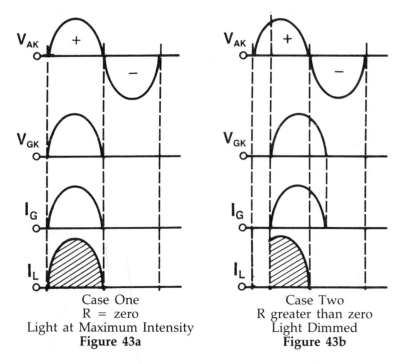

Case One
R = zero
Light at Maximum Intensity
**Figure 43a**

Case Two
R greater than zero
Light Dimmed
**Figure 43b**

With the anode voltage positive and gate current present, the SCR will conduct during the entire positive half-cycle (180°). This is the maximum conducting angle of the SCR, providing maximum average anode or load current. This condition provides the maximum achievable light intensity with the use of a single SCR. See Figure 43a.

- As the anode and gate voltages go through zero, or during the negative half-cycle, the SCR is turned OFF.

- As the A.C. voltage goes through zero from the negative to the positive region, load current resumes its flow at a point just slightly above zero as the capacitor voltage increases.

Although load current is zero during the negative half-cycle, there is a thermal inertia in an incandescent lamp that will keep its filaments in an incandescent (glowing) state even when no current is flowing and the incandescent lamp is perceived as still being ON.

If a fluorescent light bulb were to be used as the load in this circuit, an annoying ON/OFF flicker would result, since the ionized gas of the fluorescent light has no thermal inertia. Because of this flicker effect, light dimming circuits are only used with incandescent lamps.

CASE TWO (with R greater than zero):

- When the variable resistor, R, is set to some value greater than zero, the A.C. voltage is supplied to the capacitor at a later point than in Case One. Gate current is then achieved at some later time than in Case One. See Figure 43b.

- The variable resistor, R, can be adjusted to any desired value to vary the conduction angle of the load current. As the conduction angle decreases, the average load current decreases and the light is dimmed.

Theoretically, when R = ∞, the conducting angle will begin at the 180° point of the A.C. voltage, resulting in zero load current (conduction angle = zero). If the largest value of the variable resistor is equal to infinity, adjustment at its low end is very critical (poor resolution). In a practical circuit, the largest value for the variable resistor is some finite value considerably less than infinity, providing effective control at low resistance values.

The degree of light dimming is a function of the setting of the variable resistor, R, which, in turn, determines the angle of conduction of load current.

- More resistance causes a longer time for the A.C. supply voltage to be applied to the capacitor resulting in a greater degree of dimming.

- Less resistance causes a shorter time for the A.C. voltage to be applied to the capacitor resulting in a lesser degree of light dimming.

## HEATING CONTROL CIRCUIT MECHANISM

If a heating element were to be used as the load in the circuit of Figure 42, this same SCR operating mechanism would apply to achieve heating control.

At R = 0, the conduction angle is equal to 180° providing the maximum heat with the use of a single SCR. As the value of R is increased, the conduction angle is decreased and less current flows in the heating element, reducing heat output.

A heating element also has a thermal inertia. During the OFF section of the A.C. voltage (negative half-cycle), the heating element is still emitting heat.

## IMPROVING THYRISTOR EFFICIENCY

The circuit of Figure 42 uses only one SCR to achieve either lighting or heating control, however, only the positive half-cycles of the A.C. supply voltage are being used. This produces a lower average current in the load when R is equal to zero than if both half-cycles of the A.C. supply voltage were used. As a result, there is a failure in achieving the maximum intensity of the light bulb(s) and maximum heat output of the heating element.

If the circuit is modified to allow the flow of current in the load during both half-cycles of the A.C. supply voltage, the average load current would be doubled, increasing light output and heat emission. This can be accomplished by changing the circuit shown in Figure 42 so that two SCRs are connected in an inverse, parallel configuration. See Figure 44.

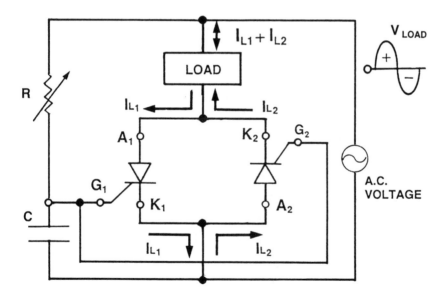

Lighting and Heating Control Circuit Using Two SCRs
**Figure 44**

As before, turn-on of the load will still be controlled by the simultaneous presence of gate current and positive voltage at the anode. As seen in the circuit of Figure 44, there are two anodes, two cathodes and a common gate. In this configuration, current will flow through the load during both half-cycles of the A.C. supply voltage.

During the positive half-cycle, SCR$_1$ is ON since its anode voltage is positive and its gate current is flowing. Current is flowing through the load during the positive half-cycle.

During the negative half-cycle, SCR$_2$ is ON since its anode voltage is positive and its gate current is flowing. Current is flowing through the load during the negative half-cycle.

## TRIAC (TRIODE A.C. SWITCH)

Instead of using two separate SCRs, the same circuit shown in Figure 44 can be operated with the use of a single triode A.C. switch, called TRIAC. This device is designed to function as two SCRs connected in an inverse, parallel configuration. The schematic symbol of the TRIAC is shown in Figure 45.

Each half of the output section of the TRIAC acts as both anode and cathode interchangeably as the A.C. supply voltage alternates, therefore, the symmetrical output terminals are now designated as Main Terminal 1 (MT1) and as Main Terminal 2 (MT2). The common gate still acts as the control terminal, since it is connected internally to both input sections of the device.

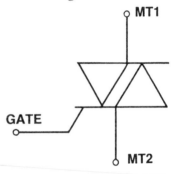

TRIAC Schematic Symbol
**Figure 45**

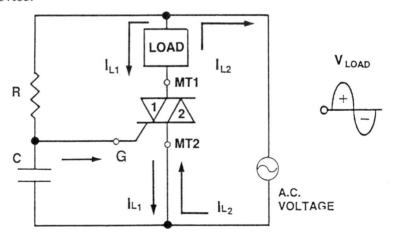

Lighting and Heating Control Using a TRIAC
**Figure 46**

The circuit of Figure 46 uses a TRIAC instead of two SCRs.

- During the positive half of the A.C. voltage, with gate current turning the TRIAC ON, current flows from the top side of the A.C. supply to one side of the load, to terminal MT1 of the TRIAC (acting as the anode of Section 1). It then flows through Section 1 to terminal MT2 of the TRIAC (acting as the cathode of Section 1) and from there to the bottom side of the A.C. voltage supply.

- During the negative half of the A.C. voltage, with gate current turning the TRIAC ON, current flows from the bottom side of the A.C. supply, to terminal MT2 of the TRIAC (now acting as the anode of Section 2). It then flows through Section 2 to terminal MT1 of the TRIAC (now acting as the cathode of Section 2) to the bottom side of the load, through the load to the top side of the A.C. voltage supply.

## TRIAC SPECIFICATIONS

The ratings and electrical characteristics of the TRIAC are essentially the same as those of the SCR. The reverse current, peak inverse, or D.C. blocking voltage specifications are imposed during the OFF portions of the A.C. voltage cycle.

## DIAC CHARACTERISTICS

A bi-directional diode A.C. switch, or DIAC, may be used in the gate circuit of a TRIAC to act as its trigger. The DIAC is a three-layer (NPN) structure with two undesignated terminals. See Figure 47.

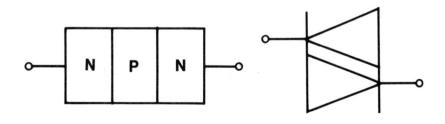

DIAC Cross-section          DIAC Schematic Symbol

**Figure 47**

The DIAC has similar switching characteristics when a voltage is applied, regardless of its polarity. See Figure 48.

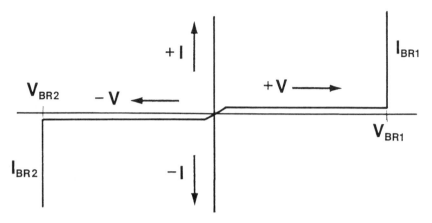

DIAC Volt-Ampere Characteristics
**Figure 48**

Because of its symmetrical characteristics, the DIAC acts as an A.C. switch that goes into its conducting state on reaching the breakover voltage ($V_{BR}$) value for either polarity, with $V_{BR1} = V_{BR2}$.

When supplied as a separate component, the DIAC is connected to the gate terminal of the TRIAC as shown in Figure 49, however, many TRIACs have their own DIAC internally connected as part of the gate circuit.

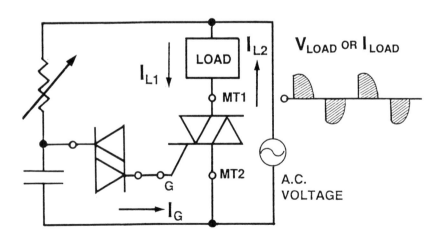

TRIAC Control Circuit with DIAC
**Figure 49**

# DIAC MECHANISM

- When the voltage across the capacitor reaches the breakover voltage of the DIAC ($V_{BR}$), the resistance of the DIAC is reduced to zero, allowing gate current to flow in the TRIAC.

- Depending on the setting of the variable resistor, R, and the $V_{BR}$ of the DIAC, the positive half-cycle of the A.C. voltage will cause gate current to flow in the TRIAC to provide a specific conduction angle. The TRIAC will stop conducting when the A.C. voltage supply is at zero.

- When the A.C. voltage goes negative, TRIAC gate current will start at the same point as it did for the positive half-cycle to provide the same conduction angle as before. The TRIAC will stop conducting when the A.C. voltage supply is at zero.

The TRIAC is turned OFF and will remain OFF when:
- The contacts of the SPST switch in the gate circuit are broken, opening the gate circuit and removing gate current.

- The load current is reduced to zero as the A.C. sine wave voltage goes through zero.

# DIAC SPECIFICATIONS

## ABSOLUTE MAXIMUM RATINGS AT 25°C

PEAK CURRENT - $I_P$
This value is specified for a 10 microsecond duration and a 120 cycle repetition rate. A typical value is 2 amperes.

POWER DISSIPATION - $P_D$
This value is specified at 25°C and must be derated from 25°C to zero power capability at maximum operating temperature. Typical values range from 250 to 350 milliwatts.

MAXIMUM OPERATING AND STORAGE TEMPERATURES
Military and aerospace applications: from -55°C to +125°C.
For consumer, commercial and industrial applications there are no standard limits. The temperature range is determined by the manufacturer and is typically -10°C to +100°C.

## ELECTRICAL CHARACTERISTICS

BREAKOVER VOLTAGE - $V_{BR1}$ AND $V_{R2}$
Generally specified at minimum, maximum and typical values.
A representative breakover voltage value is about 35 volts.

BREAKOVER VOLTAGE TEMPERATURE COEFFICIENT
Typically, this value is about 0.1% (1000 parts per million)
change per °C change in temperature.

BREAKOVER CURRENT - $I_{BR1}$ AND $I_{BR2}$
Listed as a maximum value at a specified temperature.
A typical maximum value is 200 microamperes at 25°C.

BREAKOVER VOLTAGE SYMMETRY
This is listed as $V_{BR1} = V_{BR2}$ at a specified tolerance.
A typical tolerance is plus or minus 10%.

## PACKAGE OUTLINE
The package drawing appears on the data sheet and will include
information on package and lead dimensions and material.

# ADDITIONAL SCR APPLICATIONS

## D.C. POWER SUPPLY MONITOR

The circuit shown in Figure 50 can be used to provide a visual
indication of momentary interruptions of a D.C. voltage supply.

The circuit operates as follows:
- Initially closing the normally-open contacts of the momentary
  SPST pushbutton start switch, S, applies a forward bias volt-
  age to the gate of the SCR, causing gate current to flow.
  Resistor $R_1$ is used to limit the current in the gate circuit.

- Since a positive voltage is already applied to the anode of the
  SCR, the SCR is turned ON. Because the D.C. supply voltage
  (in the form of steady-state D.C.) is applied between the
  anode and cathode of the SCR, the SCR is in a "latched" state
  and will remain ON. Resistor $R_3$ is used as the SCR load
  resistor, limiting the current in its anode-to cathode section. $R_3$
  is also used to limit the current in the indicator light circuit.

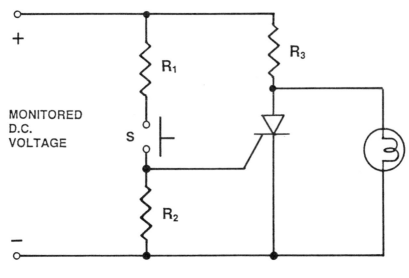

D.C. Power Supply Monitor
**Figure 50**

- Turning ON the SCR reduces its anode-to-cathode resistance to essentially zero, shorting out and turning OFF the indicator light in parallel with the anode-to-cathode circuit.

As long as the D.C. supply voltage is present, the SCR will remain ON causing the indicator light to remain OFF.

- If there is a momentary interruption of the D.C. voltage supply, generally longer than 50 microseconds, the SCR will turn OFF and remain OFF. Because of the momentary feature of switch S, its contacts will open after the initial closure.

- When the D.C. supply voltage is reapplied, the indicator light will go on, warning that service has been interrupted. Resistor $R_2$ provides a low resistance path across the gate of the SCR, preventing spurious voltages from inadvertently firing the SCR.

The only way the indicator light will turn OFF after the D.C. supply voltage is reapplied is to close the contacts of the start switch, S. The indicator light will continue to glow until the start switch is reset manually.

This circuit is used only to indicate momentary D.C. supply voltage failure.

## TRANSIENT VOLTAGE PROTECTION

A useful circuit for protection against voltage transients is shown in Figure 51. It is normally called a "crow-bar" circuit.

Under normal conditions, the two SCRs connected in opposing polarity across the A.C. voltage supply have no effect on the circuit they are protecting.

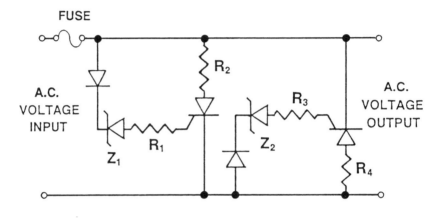

Voltage Transient Protection Circuit
**Figure 51**

- If a voltage transient in excess of the zener diode voltage occurs in either gate circuit, gate current is supplied to whichever SCR has its anode positive at that instant. The resistor ($R_1$ or $R_3$) in series with the zener diode ($Z_1$ or $Z_2$) is used to limit the current in the gate circuit of the appropriate SCR.

- This SCR starts conducting for the rest of the half-cycle, during which time the resistance between its anode and cathode is extremely small. The resistor in series with the anode ($R_2$ or $R_4$) acts to limit the current through the fuse and the anode-to-cathode circuit.

- Because the resistor in series with the anode ($R_2$ or $R_4$) is small in value, the current through the fuse is very high, causing the fuse to melt. By removing the A.C. voltage, the equipment is protected from the excess voltage transient.

- This "boot strap" circuit requires sufficiently large SCRs that are capable of sustaining high currents for several milliseconds until the fuse has blown.

# HEAT TRANSFER CONSIDERATIONS

Successful use of thyristors depends on adequate cooling to maintain operation within their rated temperature range. If the internal temperatures of these devices exceed their rated values, their electrical characteristics might be permanently changed and catastrophic failure might result.

A thyristor with a power capability greater than 1 watt at 25°C is generally packaged in a case that has a large flat surface to be clamped to a metal heat sink. Heat sinks are designed in a great variety of forms. Their purpose is to transfer the heat generated in and around the thyristor to the surrounding air or another cooling medium to prevent the case temperature of the thyristor from exceeding the manufacturer's specified limits.

In lead-mounted, lower-power devices, heat can be removed by radiation and convection if sufficient air circulation is provided within the equipment enclosure. A properly designed enclosure will generally have ventilation openings as part of its structure to allow heat to escape the circuit area. The use of fans or blowers to enhance air circulation and hot air venting is a recommended, and often necessary, technique for effective heat removal, regardless of the operating power level of the devices.

If a heat sink is used to provide optimal cooling efficiency, the thyristor package must be mounted in a proper location on the heat sink with sufficient mounting pressure (torque) to insure an intimate thermal interface between package and heat sink. Care must be taken to avoid applying too much torque, since increased torque beyond a certain point no longer improves the thermal contact. Excessive torque may mechanically stress the thyristor chip and the materials soldered or brazed to the case inside the housing, possibly resulting in permanent damage to the device. Adherence to the torque specification on the data sheet is strongly recommended and a torque wrench should always be used in mounting the case to the heat sink.

Air pockets, often created in the depressions and voids between the case and heat sink, result in inefficient heat transfer, since air is a poor thermal conductor. To fill in the voids and depressions, a thin layer of thermally conductive silicone grease is applied to the two surfaces before mounting, thereby eliminating this problem. Several types of silicone grease are commercially available for this purpose.

As a heat sink material, copper can provide optimum heat transfer because of its high thermal conductivity. Since the cost of copper is relatively high, the use of other heat conducting materials, such as steel or aluminum, may provide satisfactory performance at a lower cost. Although aluminum is lighter and is easier to fabricate than steel, in a humid or corrosive environment the resulting galvanic action between an aluminum heat sink and the copper case of a device may lead to the eventual deterioration of the thermal joint and cause an increase in thermal resistance. Proper treatment, such as plating, can eliminate this problem.

Some circuit applications necessitate electrical isolation between the thyristor case and the heat sink while maintaining good thermal contact between the surfaces. The use of a thermally conductive beryllium oxide or mica insulator between the two surfaces, in conjunction with silicone grease, will provide uniform thermal contact at the interface while effectively maintaining both heat transfer and good electrical isolation between case and heat sink.

Thermal factors should be considered in the initial design of a circuit and not as an afterthought. Well-designed and correctly implemented heat transfer systems can provide minimization of weight and space, low material costs and, most important of all, reliable operation.

When operated within their specified electrical ratings and sufficiently cooled with appropriate heat transfer equipment, properly manufactured thyristors have no inherent failure mechanisms. They have essentially infinite longevity, even in the harshest of environments, providing an extremely effective electronic tool for use in power switching control circuits.

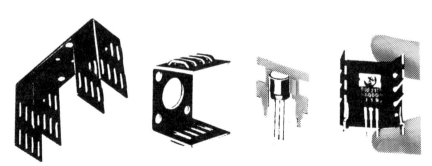

Typical Heat Sinks

# REINFORCEMENT EXERCISES

Answer TRUE or FALSE to each of the following statements:

1. The uni-directional silicon controlled rectifier (SCR) and the bi-directional triode A.C. switch (TRIAC) are in the thyristor family and are used in power control applications, such as light dimming, heat control and motor speed control.

2. Turning the SCR ON requires the simultaneous application of a positive voltage at the anode with respect to the cathode and the existence of gate current in the input circuit. If only one condition exists, the device will not turn ON and there will be no current flowing in the load.

3. When the SCR is turned ON, it will latch-up and remain ON until the gate current is reduced to zero.

4. The SCR is turned OFF by reducing the load current to zero. It will remain OFF, even if gate current is flowing, as long as the anode voltage is zero.

5. The SCR will remain OFF if the gate current stays at zero, but will turn ON when positive voltage is reapplied to the anode.

6. To reduce the load current to zero, the load circuit can be physically opened by breaking the contacts of a switch in the load circuit or by reducing the anode supply voltage to zero.

7. Since the SCR is sometimes referred to as an A.C. switch, only A.C. can be used as the supply voltage for this device.

8. The SCR is an ideal component for use in lighting and heating control circuits and other power switching applications.

9. The bi-directional triode A.C. switch (TRIAC) uses A.C. as the source of voltage in the load circuit, but requires a separate D.C. voltage source for the gate circuit.

10. TRIAC and SCR terminals are the same. The output terminal is the Anode, and the input, or control terminal, the Gate. The terminal common to both input and output sections is the Cathode.

11. The structure of a TRIAC is effectively made up of two SCRs connected in an inverse, parallel configuration allowing load current to flow during both half-cycles of the A.C. supply voltage. The A.C. load current flows alternately through one or the other of the equivalent SCR devices as the A.C. supply voltage is alternating.

12. The bi-directional diode A.C. switch (DIAC) is a 2-terminal device used in the gate circuit of a TRIAC to provide the trigger action to turn the TRIAC ON. The DIAC may or may not be part of the TRIAC.

13. Successful use of a thyristor depends on adequate cooling. If its rated temperature is exceeded, its characteristics may be permanently changed and catastrophic failure of the device may result.

14. The tighter the pressure between the thyristor case and the surface of a heat sink, the lower the thermal resistance between the two surfaces. To optimize heat transfer, as much torque as possible should be applied in mounting the case of the thyristor to the heat sink.

Answers to the reinforcement exercises are on page 192.

CHAPTER
SIX

# BIPOLAR TRANSISTORS

# BIPOLAR TRANSISTORS

## DEFINITIONS AND DESCRIPTIONS

A bipolar transistor is a 3-element semiconductor component that can function in a circuit as a linear amplifier or as a digital switch having amplification capability.

A bipolar transistor is constructed by diffusing an additional N or P region onto a P-N junction, creating either an NPN or PNP 3-layer structure. The terminals connected to each of the three regions are designated as *collector, base* and *emitter*. See Fig. 52.
- The collector is the output terminal.
- The base is the control or input terminal.
- The emitter is the reference terminal and is common to both input and output.

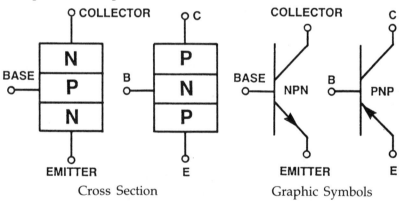

Cross Section                    Graphic Symbols
NPN and PNP Bipolar Transistors
**Figure 52**

Bipolar transistors are manufactured with two polarities, NPN and PNP, one being the complement of the other. Together, they offer the following circuit flexibilities:
- They can use either positive or negative power supplies.
- They respond to either positive or negative input voltages.
- When both types are connected in a circuit in a complementary configuration, a minimum number of parts are used.

The proper voltage polarities for the NPN and PNP transistor are shown in Figures 53a and 53b. The collector and base terminals for either type are connected to voltage supplies having the same polarity with respect to the common emitter terminal—positive for the NPN and negative for the PNP.

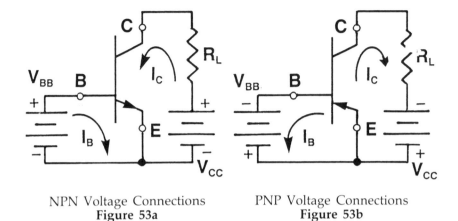

NPN Voltage Connections
**Figure 53a**

PNP Voltage Connections
**Figure 53b**

An internal base resistor, part of the device geometry, limits the flow of current in the base circuit.

NPN TRANSISTOR (Figure 53a)
Current always flows from the positive terminal of a voltage source. In an NPN transistor, collector current, $I_C$, goes from the positive terminal of the supply voltage, $V_{CC}$, flows through the load resistor, $R_L$, to the collector terminal, through the collector-to-emitter path to the emitter terminal. It then flows to the negative terminal of the collector supply voltage, $V_{CC}$, completing the output circuit.

Base current, $I_B$, leaves the positive terminal of the base supply voltage, $V_{BB}$, flows to the base terminal, through the base-to-emitter path. It then flows to the negative terminal of the base supply voltage, completing the input circuit.

PNP TRANSISTOR (Figure 53b)
The PNP transistor, the complement of the NPN, requires voltage polarities opposite to those applied to the NPN transistor. With the polarity of the supply voltages now reversed, the current flow in both output and input circuits is also reversed.

The arrowhead, which is the graphic symbol of the emitter, points in the direction of current flow. It indicates the proper voltage polarities applied to the input and output terminals of the NPN and PNP transistors.

# TRANSISTOR MECHANISM
## AS A LINEAR AMPLIFIER

Either an NPN or PNP type can be used as a model to explain the transistor mechanism when operating as part of a linear amplifier. In the following explanation, the NPN type is used. The identical explanation can be applied to the PNP type with voltage polarities reversed.

> Linear amplification is the action of a circuit that causes a small voltage at its input to change into a larger voltage at its output with no change in its appearance or waveshape.

In an amplifier circuit, the ratio of output voltage, $V_{OUT}$, to input voltage, $V_{IN}$, is called the *voltage gain* of the circuit.

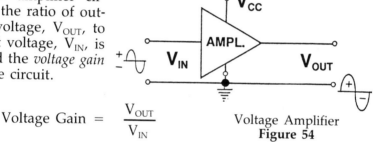

$$\text{Voltage Gain} = \frac{V_{OUT}}{V_{IN}}$$

Voltage Amplifier
**Figure 54**

The NPN bipolar transistor is shown in cross-sectional form in Figure 55. Initially, the input (base-to-emitter) is open with no connection to the base terminal. The output circuit consists of a D.C. supply voltage, $V_{CC}$, with its positive terminal connected through a load, $R_L$, to the collector terminal. The negative terminal of $V_{CC}$ is connected to the emitter terminal.

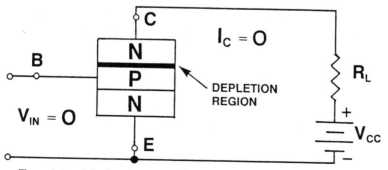

Transistor Mechanism as a Linear Amplifier - Initial Step
**Figure 55**

Under this condition, the collector-to-base NP junction is in a reverse biased state (positive to the N-region), creating an infinite resistance depletion layer between these regions. Since the collector-to-emitter resistance, $R_{CE}$, is infinite, no current will flow in the circuit.

If a base supply voltage, $V_{BB}$, is now connected between base and emitter (positive to the base), the base-to-emitter PN junction is *forward biased* and base current, $I_B$, will flow.

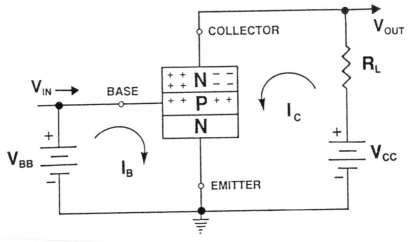

Transistor Mechanism as a Linear Amplifier - Final Step
**Figure 56**

The flow of base current, limited by the internal resistance in the base, $R_{BE}$, produces positive ions in the P-region. The excess positive ions migrate through the depletion region to the upper N-region, causing some of the upper N-material to invert to P-material.

The partial inversion of the upper N-region changes the initial collector-to-emitter resistance, $R_{CE}$, from an infinite value to a desired finite value of resistance.

Changing $R_{CE}$ from infinity to a specific value produces a two-element series circuit, $R_L + R_{CE}$ connected across the supply voltage, $V_{CC}$. With the output circuit now closed, output current, $I_C$, will flow. The path of the output current is from the positive terminal of the supply voltage, $V_{CC}$, through the load resistor, $R_L$, to the collector terminal, through the collector-to-emitter resistance, $R_{CE}$, to the emitter terminal and then to the negative terminal of the supply voltage, $V_{CC}$.

## CURRENT GAIN AND VOLTAGE GAIN

Associated with every bipolar transistor is a characteristic called *current gain* or *Beta* ($\beta$), the ratio of output current, $I_C$, to input current, $I_B$. It is expressed mathematically as:

$$\text{Current Gain} = \text{Beta} \ (\beta) = \frac{I_C}{I_B}$$

The value of the collector current, $I_C$, will depend on the value of Beta and the value of the base current, $I_B$:

$$I_C = I_B \times \beta$$

The collector current, $I_C$, is supplied from the voltage source, $V_{CC}$. In flowing through the load resistor, $R_L$, the collector current produces a voltage across the load resistor, called the output voltage, $V_{OUT}$, equal to the product of the collector current times the load resistor:

$$V_{OUT} = I_C \times R_L$$

If the initial forward bias voltage, $V_{BB}$, and resulting base current, $I_B$, is varied by applying a changing input signal, $V_{IN}$, between base and emitter, the output current, $I_C$, will vary accordingly.

- As base current increases, collector current increases.
- As base current decreases, collector current decreases.

To identify a changing condition, the Greek symbol delta ($\Delta$) is used before the parameter designation.  For example, a changing input (base) voltage is shown as $\Delta V_{IN}$ and a changing output (collector) current is shown as $\Delta I_C$.

The changing input current, $\Delta I_B$, produces a changing output current, $\Delta I_C$ ($\Delta I_B \times \beta$). This changing current, in turn, creates a changing output voltage, $\Delta V_{OUT}$, since $\Delta V_{OUT} = \Delta I_C \times R_L$.

The ratio of a changing output voltage, $\Delta V_{OUT}$, to a changing input voltage, $\Delta V_{IN}$, is the voltage gain of an amplifier.

$$\text{Voltage Gain} = \frac{\Delta V_{OUT}}{\Delta V_{IN}}$$

To set the transistor into the proper operating mode for linear amplification before an input signal is applied, the correct initial value of collector current must be established. This is called the *quiescent collector current* ($I_{C(Q)}$). It is fixed by the initial value of $I_B$ and beta. The appropriate value of $I_{C(Q)}$ is generally set approximately midway between the upper and lower limits of the collector current.

Linear amplification means that the amplified output voltage is an accurate reproduction of the input voltage. The distance between the upper and lower limits of the output current is called the dynamic range of the output circuit. Exceeding the *dynamic range* (too little or too much collector current) in either direction will result in an inaccurate, or distorted output voltage. Generally, a maximum value of distortion is listed in the specification of a linear amplifier.

## REVIEW - TRANSISTOR MECHANISM AS A LINEAR AMPLIFIER

An equivalent of a linear amplifier is shown in Figure 57.

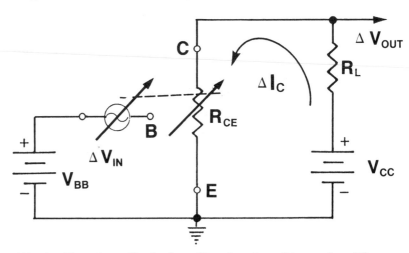

Bipolar Transistor Equivalent Circuit - As a Linear Amplifier
**Figure 57**

After the initial forward bias has set the proper value for $I_{C(Q)}$, the varying input $\Delta V_{IN}$, will cause the collector-to-emitter resistance, $R_{CE}$, to act as a variable resistor. The variation of $R_{CE}$ will vary $\Delta I_C$, which, in turn, creates the varying output voltage, $\Delta V_{OUT}$ ($\Delta I_C \times R_L$). The voltage gain of the amplifier is equal to $\Delta V_{OUT}$ divided by $\Delta V_{IN}$.

## TRANSISTOR MECHANISM AS A DIGITAL SWITCH

Either an NPN or PNP type can be used to explain the transistor mechanism when used as part of a digitally-controlled switch. In the following explanation, the NPN type is used. The identical explanation can be used with the PNP type with voltage polarities reversed.

When an instantaneous increase in base current (step-function) is applied to the input terminals of a digital switch, it will cause the circuit to suddenly change from its OFF to its ON state. The output section of the circuit will revert to its initial OFF state when the base current is removed.

### INITIAL STEP - SWITCH OFF
The NPN bipolar transistor is shown in cross-sectional form in Figure 58. Initially, the input voltage, $V_{IN}$, applied between base and emitter is zero. The supply voltage, $V_{CC}$, is connected between the collector and emitter terminals through load resistor, $R_L$, with positive polarity to the collector.

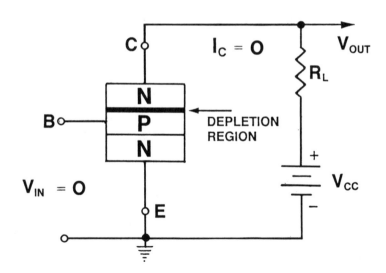

Digitally-Controlled Switch Mechanism - Initial Step
**Figure 58**

Under this circuit condition, the collector-to-base NP junction is reverse biased (positive voltage to the N-region) creating an infinite resistance depletion layer between the N-region and P-region. Since collector-to-emitter resistance, $R_{CE}$, is infinity, no output current is flowing ($I_C = 0$).

SECOND STEP - TURNING THE SWITCH ON
- If a positive step voltage, $V_{IN}$, is now applied between base and emitter, the base-to-emitter PN junction is forward biased and base current, $I_B$, will flow. See Figure 59. The magnitude of the input voltage must be high enough to produce a sufficiently high base current so that all of the upper N-region inverts to P-type material, thereby creating a forward biased PN junction between collector and emitter.

- With the upper N-region totally inverted to P-material, the transistor is now in a state of *saturation* and resembles a forward biased diode. Saturation is that condition where any further increase in base current will cause no further decrease in collector-to-emitter resistance.

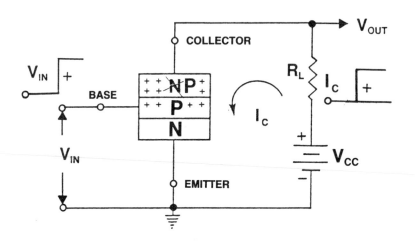

Digitally-Controlled Switch Mechanism - Second Step
**Figure 59**

- The collector-to-emitter resistance, $R_{CE}$, is reduced to its lowest possible value (essentially zero). The output circuit is closed, allowing collector current to flow.

The condition of transistor saturation is specified either as collector saturation resistance, $R_{CE(sat)}$, or as collector saturation voltage, $V_{CE(sat)} = (I_C \times R_{CE(sat)})$. In saturation, the following relationships exist:

$$I_C = \frac{V_{CC} - V_{CE(sat)}}{R_L} \quad \text{and} \quad V_{OUT} = V_{CC} - I_C \times R_L = V_{CE(sat)}$$

REVERTING TO THE INITIAL STEP -
TURNING THE SWITCH OFF

When the input voltage returns to zero, $R_{CE}$ reverts to infinity, $I_C$ = zero and $V_{OUT}$ = $V_{CC}$.

The transistor operates very efficiently (very low power losses) when used as a digital switch. Most of the power is generated in the load resistor ($I_C^2 \times R_L$) with very little power dissipated in the transistor.

- When the transistor is OFF, there is no input or output current and no power is being dissipated.

- When the transistor is ON, base current is relatively low and very little power is dissipated in the base circuit. With $R_{CE(sat)}$ essentially at zero, even with relatively high collector current, the voltage between collector and emitter is essentially zero, therefore, essentially no power is being dissipated between collector and emitter.

REVIEW - TRANSISTOR MECHANISM
AS A DIGITAL SWITCH

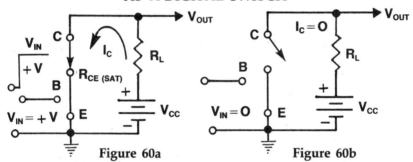

**Figure 60a**                    **Figure 60b**

Bipolar Transistor Equivalent Circuit - As a Digital Switch

- A positive step input voltage, $V_{IN}$, produces sufficiently high forward bias to saturate the transistor. The initial collector-to-emitter resistance changes from infinity (open circuit) to zero ($R_{CE(sat)}$). See Figure 60a. This causes the output circuit to switch ON, allowing collector current to flow. The output voltage, $V_{OUT}$, is between the collector and emitter (reference) and is essentially equal to zero.

$$V_{OUT} = V_{CC} - (I_C \times R_L) = V_{CE(sat)}$$

- When the input voltage returns to zero, $R_{CE}$ reverts to its initial value of infinity and the transistor switches OFF. At that time, $I_C$ = zero and $V_{OUT}$ = $V_{CC}$. See Figure 60b.

# BIPOLAR TRANSISTOR SPECIFICATIONS

To assist in the selection of a proper device for a particular application, data sheet information lists the device polarity, its material and the manufacturing process. Included are its physical outline and dimensions, its significant electrical features and its general and/or unique applications.

POLARITY
NPN - positive voltage on both collector and base.
PNP - negative voltage on both collector and base.

Selection of the transistor polarity is based on:
• The polarity of the supply voltage.
• The polarity of the D.C. signal to be amplified.
• The polarity of the switching voltage used to control the transistor being used as a digital switch.

The major specifications are classified under the headings of *absolute maximum ratings* and *electrical characteristics*.

Absolute maximum ratings are the electrical, mechanical and thermal limits of the device assigned by the manufacturer. If these ratings are exceeded, permanent damage to the device may result. These values do not specify the actual characteristics of the transistor under operating conditions.

Electrical characteristics are the measurable electrical properties of the device inherent in its design and are listed at a specific operating temperature. The characteristics describe the features of the device under conditions which approximate its actual operation.

## ABSOLUTE MAXIMUM RATINGS -
## BIPOLAR TRANSISTORS

Voltage ratings specify breakdown values as follows:

• With reverse voltage applied to a selected junction.

• With voltage applied across two junctions - with one junction reverse biased and the second junction in some specified state.

Single junction breakdown from collector-to-base, or between emitter and base, is the same as that specified for a reverse biased diode.

## COLLECTOR-TO-BASE BREAKDOWN VOLTAGE - $BV_{CBO}$
Maximum allowable reverse bias D.C. voltage that may be applied between collector and base with an open emitter. Depending on the device, this value will usually range between 25 and 150 volts.

## BASE-TO-EMITTER BREAKDOWN VOLTAGE - $BV_{BEO}$
Maximum allowable reverse bias voltage that may be applied between base and emitter with an open collector. Depending on the device, this value will usually range between 2 and 25 volts.

## COLLECTOR-TO-EMITTER BREAKDOWN VOLTAGE - $BV_{CE}$
Maximum allowable D.C. voltage applied between collector and emitter is specified under various conditions of the base terminal with respect to the emitter as follows:

$BV_{CEO}$ - with the base open—the most demanding condition. Depending on the device, this value will usually range between 12 and 150 volts and indicates the maximum allowable magnitude of the D.C. supply voltage.

$BV_{CER}$ - with a resistor connected between base and emitter— a less stringent condition.

$BV_{CES}$ - with the base shorted to the emitter—the next less stringent condition.

$BV_{CEX}$ - with a reverse bias applied to the base-to-emitter junction—the highest voltage rating condition.

For all of the breakdown voltage rating designations, the letter "B" which precedes the letter "V" is often omitted, e.g. $V_{CBO}$, $V_{CEO}$, $V_{CER}$, etc. is the same as $BV_{CBO}$, etc.

## COLLECTOR CURRENT, D.C. - $I_{Cmax}$
Maximum allowable D.C. collector current. This value will depend on the size of the chip and package. For power devices, it is typically in the ampere range. For small-signal devices, the range of values is in milliamperes.

## COLLECTOR CURRENT, PEAK - $I_{Cpeak}$
Maximum pulse current rating for a specified pulse width. When operating as a digital switch with very short, high current pulses, the power dissipated in the transition from OFF to ON must be taken into account to assure that the ratings of the device are not exceeded. Depending on the device, this rating can range from milliamperes to amperes.

BASE CURRENT, D.C. - $I_{Bmax}$
Maximum allowable steady-state D.C. base current. For power devices, typical values are in milliamperes. For small-signal devices, typical values are in microamperes.

BASE CURRENT, PEAK - $I_{Bpeak}$
Maximum allowable base pulse current rating for a specified pulse width. Depending on the device, this rating ranges from microamperes to milliamperes.

POWER DISSIPATION - $P_D$
Maximum allowable power dissipation at 25°C case temperature. This assumes the use of an infinite heat sink to maintain the case temperature at 25°C. This rating is a reference value and is derated with increasing temperature as shown below.

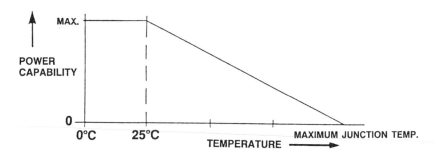

Power Dissipation Derating Curve
**Figure 61**

*Power derating factor* is defined as the decrease in power capability as a function of increasing temperature. It is measured in watts per °C under specified conditions of heat transfer. Maximum power rating is generally listed at 25°C with no power increase allowed at a lower temperature.

Refer to CHAPTER FIVE — THYRISTORS for a discussion of heat sinking considerations for power devices.

OPERATING AND STORAGE TEMPERATURE RANGE
For military and aerospace applications: -55°C to +125°C. For commercial, industrial and consumer applications, there are no standard limits. The temperature range is specified by the manufacturer and typically extends from -10°C to +100°C.

## ELECTRICAL CHARACTERISTICS - BIPOLAR TRANSISTORS

Electrical characteristics are listed at a temperature of 25°C unless otherwise noted. In addition to typical values, minimum and maximum values are given where applicable.

### COLLECTOR CURRENT CUTOFF (LEAKAGE) - $I_{CES}$

This characteristic specifies the maximum and typical leakage current in the collector circuit under the condition that the base is shorted to the emitter at a specified $V_{CC}$. In many cases, values for this characteristic are given at elevated temperatures to indicate the quality of a transistor as a switch that is OFF under high temperature conditions. The value of $I_{CES}$ approximately doubles for every 10°C rise in temperature.

- For small-signal silicon transistors, at 25°C, typical values range between 0.1 to 10 nanoamperes.
- For power devices, the range is from 0.1 to 1 milliampere.

### CURRENT GAIN (FORWARD CURRENT TRANSFER RATIO)- Beta (β) or $h_{FE}$

This characteristic specifies the ratio of collector current, $I_C$, to base current, $I_B$, (also $\Delta I_C / \Delta I_B$), at a specified value of collector current and supply voltage, $V_{CC}$. The emitter terminal is common to both input and output.

A typical beta curve is shown in Figure 62. Note that as collector current increases from zero, beta increases. It levels off at some value of collector current and then decreases as collector current increases beyond this point. To prevent an increase in junction temperature as collector current is increased, these measurements are made in a short period for a few microseconds.

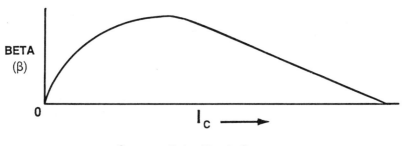

Current Gain (Beta) Curve
**Figure 62**

## FREQUENCY CUTOFF - $F_{CO}$

This characteristic indicates the frequency capability of the transistor when operated as an amplifier. As the frequency of the input signal is increased, the current gain ($\beta$) of the device at a fixed value of collector current is constant and then starts to fall off. The point at which the beta drops to 0.707 of the initial lower frequency beta is the frequency cutoff of the device. When using the transistor as an amplifier, the input signal frequency should be limited to a value below its frequency cutoff value. Frequency response can be increased by operating at a high value of collector current to obtain the highest beta. A typical frequency response curve is shown in Figure 63.

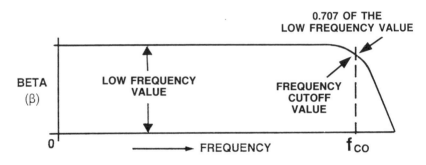

Frequency Response Curve
**Figure 63**

## COLLECTOR SATURATION RESISTANCE - $R_{CE(sat)}$

This characteristic specifies the maximum and typical values of collector-to-emitter resistance which occur at the point where an increase in base current no longer causes a further decrease in collector-to-emitter resistance. This is called collector saturation. Prior to collector saturation, if base current is increased, the collector-to-emitter resistance decreases.

When the transistor is operated as a switch, the input step voltage provides sufficiently high base forward bias voltage (and base current) to drive the transistor into saturation.

- With a small-signal transistor having a typical value of $R_{CE(sat)}$ of about 5 $\Omega$, an $R_L$ of about 5000 $\Omega$ might be used.
- With a power transistor having a typical $R_{CE(sat)}$ value of about .001 $\Omega$, a value for $R_L$ of about 20 $\Omega$ might be used.

Typical switching circuits are designed so that $R_{CE(sat)}$ is negligible compared with $R_L$.

## COLLECTOR SATURATION VOLTAGE - $V_{CE(sat)}$

This characteristic is the product of collector current, $I_C$, and collector saturation resistance, $R_{CE(sat)}$. Either one or the other value is specified on a manufacturer's data sheet.

## TURN-ON TIME - $T_{ON}$

When driven into saturation, there is a time required for the collector to respond to an input digital pulse applied to the base, including both *delay time* ($t_d$) and *rise time* ($t_r$).

- Delay time is the interval between the start of the input pulse and the time it takes to reach 10% of the final $I_C$.
- Rise time is the interval between the 10% and 90% points of the final collector current. See Figure 64.

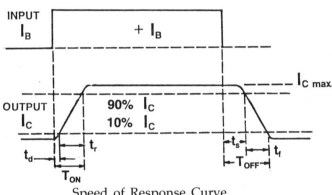

Speed of Response Curve
**Figure 64**

## TURN-OFF TIME - $T_{OFF}$

When the input pulse returns to zero, there is a time interval for the saturated transistor to turn OFF, including both *storage time* ($t_s$) and *fall time* ($t_f$).

- Storage time is the interval between the return of the input pulse to zero and the time it takes to reach 90% of the collector current.
- Fall time is the interval between the 90% and 10% points of the collector current. See Figure 64.

The geometry and size of the transistor chip will determine the speed of the device. Small-signal devices are generally very fast and power devices relatively slow. If the turn-on and turn-off times are 50 nanoseconds or less, the device is classified as a computer switch. If these times are longer, it is specified as a general-purpose switch.

# BIPOLAR TRANSISTOR APPLICATIONS

Bipolar transistors are the active components in current and voltage amplifers, at D.C. and all A.C. frequencies, and at all small-signal and power levels. Their amplification ability is used in a multitude of other applications including:

- timing circuits
- pulse generators
- sine-wave generators (oscillators)
- square-wave generators

When used in a variety of regulated power supplies, transistors provide the required regulating capability as error amplifiers, variable resistors or controlled switches in the voltage regulating section of a power supply.

In digital circuitry, the transistor is used to provide high speed switching with current gain capability. These types of circuits are necessary to satisfy the great variety of logic functions—the major sections of instrumentation systems, control systems, calculators and computers.

## AMPLIFIER CIRCUIT

The circuit shown in Figure 65 uses a bipolar transistor in an amplifier application. The current gain or beta of the bipolar transistor acts to provide the necessary voltage gain. (Refer to the discussion of the transistor mechanism as a linear amplifier covered in a previous part of this chapter.)

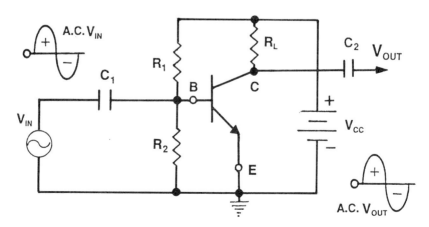

Typical Transistor Amplifier Stage
**Figure 65**

## CIRCUIT DESCRIPTION

- Instead of using a separate base voltage supply to establish the forward bias voltage for the base-to-emitter junction, a simple voltage divider network, $R_1$ and $R_2$ in series, is connected across the collector voltage supply, $V_{CC}$, with the junction of $R_1$ and $R_2$ connected to the base terminal. The value of the base forward bias voltage, $V_{BB}$, is determined by the product of $V_{CC}$ and the ratio of $R_2$ to $(R_1 + R_2)$.

$$V_{BB} = V_{CC} \times \frac{R_1}{R_1 + R_2}$$

- Capacitor, $C_1$, is used to couple (transfer) the voltage from a previous amplifier (or from any signal source such as a microphone or phonograph cartridge) to the amplifier input terminals (base-to-emitter).

- Capacitor, $C_2$, is used to couple the output voltage developed across the load resistor, $R_L$, to the next amplifier stage.

- If there are two or more amplifier stages connected in this manner, the combination is called a *cascaded* amplifier section. The gain of two cascaded amplifiers, for the most part, depends on the beta of each transistor. The beta of one transistor is multiplied by the beta of the other. The combined beta is equal to $\beta_1 \times \beta_2$. If each beta were 100, the combined beta of the cascaded transistors would be 10,000.

## DARLINGTON TRANSISTOR

If two transistors were connected in a single package configured in the manner shown in Figure 66, it is then called a *Darlington* transistor. This device has the advantages of using fewer components, providing easier assembly techniques with less interconnecting wire.

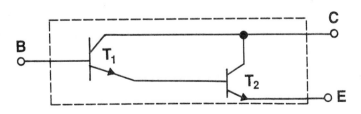

Darlington Transistor
**Figure 66**

The two devices connected in a Darlington configuration are treated as a single transistor in one package. Both collectors are connected internally and designated as one external terminal. The emitter of the first transistor is connected internally to the base of the second and designated as the external emitter terminal. The base of the first transistor is designated as the external base terminal.

Darlington transistors are manufactured as single power devices having the capability of very high beta (typically from 10,000 to over a hundred thousand) at high collector current levels (5 to 25 amperes). They are often called "super beta" transistors and are used in motor driving circuits and similar systems where a small input current is used to generate and control a relatively high load current.

## DIGITAL SWITCH WITH GAIN

Bipolar transistors are used in digital switching control and logic circuit applications as the device to simultaneously provide both switching and gain capabilities. Unlike a bipolar transistor amplifier used for linear voltage gain, a bipolar digital switch needs no initial base forward bias voltage.

At zero input signal, there is collector current and the switch is in an OFF state. With a step voltage applied to the input (base terminal), the collector is saturated and the transistor switch is turned ON.

When the input digital pulse reverts to zero, the current in the collector ceases to flow and the transistor switch is turned OFF. A typical circuit is shown in Figure 67.

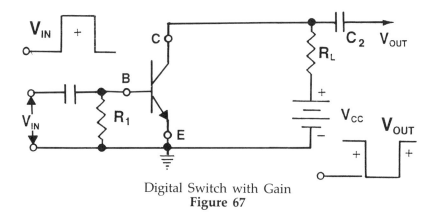

Digital Switch with Gain
**Figure 67**

Since the transistor in its OFF state is drawing essentially zero current, there is essentially zero power being dissipated. During its ON state, the collector-to-emitter resistance, $R_{CE(sat)}$, and the collector-to-emitter voltage, $V_{CE(sat)}$, are essentially zero; power dissipation is also essentially zero. When both collector-to emitter voltage and collector current are present, it is during the switching intervals from one state to the other that significant power is being dissipated. With very short switching times between ON and OFF states, very little power is dissipated. To minimize power losses, it is necessary that bipolar transistors used in high-speed switching have fast speed of response (short rise and fall times).

For computer and other logic circuitry, the turn-on and turn-off times are limited to a maximum value of 50 nanoseconds. Since these systems generally use a very large number of switching stages, the high speed requirement serves two purposes:

- Minimization of system power losses.
- Completion of a total switching sequence, or *instruction*, in a relatively short time.

## BIPOLAR TRANSISTOR CHARACTERISTICS AND TRADE-OFFS

DESIRABLE FEATURES
- Because of the nature of bipolar transistor technology, the device is capable of both current and voltage amplification. The product of voltage and current is power, therefore, it can also be used as a power amplifier.

- Some small-signal bipolar transistors are manufactured with excellent high frequency response capabilities. These types are specified to operate in the Gigahertz range and still exhibit satisfactory beta characteristics.

- The same geometric and metallurgical characteristics of bipolar transistor technology that result in high frequency response capability also produce very fast switching speed capability in high speed computer circuitry. These small-signal bipolar transistors, characterized on a data sheet for use in digital switching circuits, are capable of response times ranging from 1 to 10 nanoseconds. Improved production techniques are producing new small-signal devices capable of switching in hundreds of picoseconds.

UNDESIRABLE FEATURES

- To improve the high frequency response of a circuit using small-signal bipolar transistors, the circuit is designed with increased levels of collector current. This results in higher power dissipation (Power = $V_{CE} \times I_C$). In a multi-stage amplifier, total transistor dissipation power could be considerable. Reduction of these losses can be achieved by reducing collector current, however, its frequency response will also be reduced.

- To establish the proper collector quiescent current level of a bipolar transistor when operating as an amplifier, its input circuit (base-to-emitter junction) must be forward biased. This causes the transistor to have a low-resistance input characteristic that will adversely affect the gain of the preceding circuit by loading its output section.

- The input section of a bipolar transistor amplifier is effectively connected in parallel with the relatively higher output resistance of the preceding amplifier stage. This parallel connection will act to decrease the equivalent load resistance of the preceding amplifier stage. Since voltage gain of an amplifier is proportional to its load resistance, the voltage gain of that stage will also be reduced. The second amplifier circuit can be modified by changes in its configuration and/or with the addition of other components to increase its input resistance to avoid this loading effect. The addition of more parts, however, will increase the total cost of the circuit and take more space on a circuit board.

When used as a digital switch, power losses of a bipolar transistor are essentially zero during its ON and OFF periods. When the transistor operates at relatively high collector currents, maximum switching speeds are achieved. Power losses during the switching intervals must be examined carefully. These power losses depend on current, turn-on and turn-off times, frequency of switching and the switching duty cycle. Power losses may be reduced by lowering collector current, but this may result in reduced switching speeds.

# GLOSSARY OF SUBSCRIPT NOTATIONS FOR BIPOLAR TRANSISTORS

$BV_{BEO}$ - BASE-TO-EMITTER BREAKDOWN VOLTAGE
Maximum allowable reverse biased D.C. voltage between base and emitter with the collector open.

$BV_{CBO}$ - COLLECTOR-TO-BASE BREAKDOWN VOLTAGE
Maximum allowable reverse biased D.C. voltage between collector and base with the emitter open.

$BV_{CEO}$ - COLLECTOR-TO-EMITTER BREAKDOWN VOLTAGE
Maximum allowable D.C. voltage between collector and emitter with the base open.

$BV_{CER}$ - COLLECTOR-TO-EMITTER BREAKDOWN VOLTAGE
Maximum allowable D.C. voltage between collector and emitter with a resistor connected between base and emitter.

$BV_{CES}$ - COLLECTOR-TO-EMITTER BREAKDOWN VOLTAGE
Maximum allowable D.C. voltage between the collector and emitter with the base shorted to the emitter.

$BV_{CEX}$ - COLLECTOR-TO-EMITTER BREAKDOWN VOLTAGE
Maximum allowable D.C. voltage between the collector and emitter with reverse bias applied to the base-to-emitter junction.

$F_{CO}$ - FREQUENCY CUTOFF
The frequency at which the current gain (beta) in a linear amplifier, at a specified collector current, drops to .707 of its initial low frequency value.

$h_{FE}$ OR BETA ($\beta$) - CURRENT GAIN
Ratio of collector current to base current (also $\Delta I_C / \Delta I_B$) at a specified value of collector current and supply voltage.

$I_B max$ - BASE CURRENT, D.C.
Maximum allowable steady-state D.C. base current.

$I_{CES}$ - COLLECTOR CURRENT CUTOFF (LEAKAGE)
Maximum collector current with base shorted to emitter.

$P_D$ - POWER DISSIPATION
Maximum allowable power dissipation with the case temperature kept at 25°C by the use of a heat sink. This 25°C rating must be derated with increasing case temperatures.

## $R_{CE}$ - COLLECTOR-TO-EMITTER RESISTANCE
The general term for the collector-to-emitter resistance at a specified value of $I_B$, $V_{CE}$ and temperature.

## $R_{CE(sat)}$ - COLLECTOR SATURATION RESISTANCE
The value of collector-to-emitter resistance under the condition of collector saturation. Compared with the load resistance, $R_{CE(sat)}$ is considered to be essentially zero.

## $R_L$ - LOAD RESISTANCE
The resistance in the output circuit across which voltage is developed when output current is flowing.

## $T_{OFF}$ - TURN-OFF TIME
The interval of time required for the collector current to cut off after the input step voltage is reduced to zero.

## $T_{ON}$ - TURN-ON TIME
The interval of time required for the collector current to increase to its maximum value after a voltage step is applied to the input terminals.

## $V_{BB}$ - BASE SUPPLY VOLTAGE
For purposes of circuit explanation, this notation is used for the separate base supply voltage that provides the initial forward bias voltage for amplifier operation.

## $V_{BE}$ - BASE-TO-EMITTER VOLTAGE
The general value of voltage between base and emitter.

## $V_{CC}$ - SYSTEM SUPPLY VOLTAGE
The steady-state D.C. supply voltage for bipolar transistor circuits. If NPN transistors are used, this voltage is positive with respect to the emitter. For PNP devices, the voltage polarity is reversed.

## $V_{CE}$ - COLLECTOR-TO-EMITTER VOLTAGE
The general value of voltage between collector and emitter.

## $V_{IN}$ - INPUT VOLTAGE
The voltage, either linear or digital, applied between the input terminal of a circuit and its reference.

## $V_{OUT}$ - OUTPUT VOLTAGE
The voltage, either linear or digital, between the output terminal of a circuit and its reference.

# REINFORCEMENT EXERCISES

Answer TRUE or FALSE to each of the following statements:

1. A bipolar transistor is a three-terminal semiconductor device providing current gain between its input and output sections. Current gain of the transistor is called beta ($\beta$), defined as the ratio of the output (collector) current to its input (base) current, or as the change in collector current ($\Delta I_C/\Delta I_B$).

2. The PNP transistor is the complement of the NPN transistor, requiring opposite voltage polarities at its terminals.

3. The proper voltage polarity required by the PNP is positive at the collector terminal and positive at the base terminal. The NPN transistor requires negative voltage at its collector and negative voltage at its base.

4. A bipolar transistor is used either as an amplifying or switching active component in a circuit and can be specified on a data sheet for either application or both.

5. A bipolar transistor is specified by polarity (NPN/PNP), voltage, current and power rating, operating temperature range, leakage current, current gain and frequency response; if used as a switch, by the speed of response and saturation resistance (or saturation voltage).

6. It is not necessary to be concerned with power derating considerations for bipolar transistors since manufacturers take the increased operating temperature into account when power capability is specified on a data sheet.

7. The cutoff, or leakage current specification ($I_{CES}$) on a data sheet is an indication of the quality of the transistor as a turned-off switch.

8. The current gain or beta of a bipolar transistor increases as collector current increases and then levels off to a constant value of beta, regardless of any further increase in collector current.

9. A bipolar transistor is generally selected to be used as an amplifier if its cutoff frequency specification is the same as the highest frequency of the input signal.

10. Darlington transistors provide extremely high current gain and are used where high gain at high values of collector current is needed.

11. The power dissipation in the transistor of a single transistor amplifier circuit is considerably higher than that of the transistor in a single transistor switching circuit.

12. The desirable features of a bipolar transistor are:
    - Voltage and current (power) amplification capability.
    - For small-signal devices, high frequency response and fast switching speed.
    - *Most important of all*, very high input resistance characteristics that produce no loading of preceding circuitry.

Answers to the reinforcement exercises are on page 193.

CHAPTER
SEVEN

# FIELD-EFFECT TRANSISTORS

HISTORY OF THE
FIELD-EFFECT TRANSISTOR (FET)

FET CATEGORIES, DEFINITIONS
AND DESIGNATIONS

ADVANTAGES OF FETs
COMPARED WITH JFET AND
MOSFET CONSTRUCTION

JFET AND MOSFET MECHANISM -
LINEAR AND DIGITAL

JFET AND MOSFET SPECIFICATIONS

CMOS TECHNOLOGY

FET APPLICATIONS

REINFORCEMENT EXERCISES

# FIELD-EFFECT TRANSISTORS

A Field-Effect Transistor is a three terminal semiconductor device with an input and output section. One of its major characteristics is its very high input resistance, allowing it to function in a variety of applications without loading a preceding circuit. The effect of an electric field, produced by applying a voltage to its input, changes the resistance of its output section and gives the device its name.

## HISTORY OF THE FIELD-EFFECT TRANSISTOR (FET)

Although field-effect transistors became commercially available about eight years after bipolar transistors were in common use, the technical principle of field-effect transistors was known before the principle of the bipolar type was explored.

Considering the logical progression of component development, the field-effect transistor should have preceded the bipolar transistor as the first commercial semiconductor. The field-effect transistor (not the bipolar transistor) is truly the solid-state version of the vacuum tube triode.

The FET was first described in the technical literature in 1928. Because of a lack of interest in solid-state physics and the continuing expansion of existing vacuum tube technology during that period, field-effect transistors were nothing more than an interesting laboratory novelty.

In 1948, William Shockley of the Bell Telephone Laboratories, one of the inventors of bipolar transistors, proposed a workable field-effect device. Again, the FET was side-tracked in favor of bipolar transistor production because of the failure to achieve certain device characteristics in the FET that were favorable to practical circuitry. In the early 1960s, production problems were solved, leading to the successful manufacture of the field-effect transistors of today.

The FET has established itself as a major circuit component, having some important advantages over bipolar transistor technology, including circuit simplicity for similar applications.

# FET CATEGORIES, DEFINITIONS AND DESIGNATIONS

There are two basic types of field-effect transistors - the JFET and the MOSFET. Both operate on the principle of having an output current controlled by an electric field produced by an input voltage, however, each has its own operating mechanism. Each is manufactured in either N-type or P-type polarity.

- JFET - Junction Field-Effect Transistor
  The input section has a built-in diode (PN junction) that operates in a reverse biased mode with its input voltage controlling output current.

- MOSFET - Metal-Oxide Semiconductor Field-Effect Transistor (also called IGFET—Insulated-Gate Field-Effect Transistor) The input section has a built-in capacitor with the input voltage across the capacitor controlling output current.

Since the FET operating mechanism differs from the bipolar transistor operating mechanism, its terminals are designated differently.
- The output terminal is called the *Drain* (D).
- The input terminal is called the *Gate* (G).
- The terminal common to input and output is the *Source* (S).

The MOSFET has a fourth element, designated as the *body* or *substrate*. This element is either connected internally to the source or is available as an external terminal.

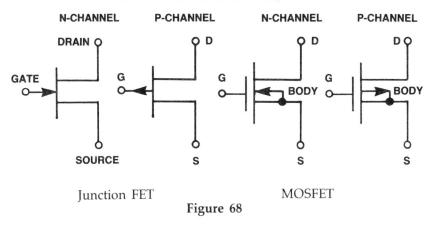

Junction FET            MOSFET

**Figure 68**

The output section (the path between drain and source) is called the *channel*. Its type (N or P) identifies the polarity of the FET.

As with bipolar transistors, junction FETs and MOSFETs are produced in complementary polarities and are designated as either N-channel or P-channel devices. FETs with complementary polarities offer the same circuit flexibilities as do the complementary bipolar transistors. The proper voltage polarities for each type are indicated in Figure 68.

The operating characteristics of the field-effect transistor are similar to those of the vacuum tube triode and yet have all the advantages inherent in semiconductor technology:
- Low power dissipation
- Low voltage requirements
- No filaments to burn out
- No warm-up time requirements
- Shock and vibration proof characteristics
- Small size
- Infinite life

While field-effect transistors and bipolar transistors can be used in similar circuitry to provide linear amplification and digital switching, there are several significant differences between them in characteristics and applications.

# ADVANTAGES OF FETS COMPARED WITH BIPOLAR TRANSISTORS

- Unlike a bipolar transistor, a FET has an extremely high input resistance (essentially infinite), a characteristic similar to that of the vacuum tube. The relatively low input resistance circuit of a bipolar transistor will load down the output circuit of a preceding amplifier stage and adversely affect its gain. In contrast, the FET has no loading effect.

- Because of its extremely high input resistance, fewer added components are used for FET circuitry. Simpler circuitry can achieve the same result as the more complex circuits required with bipolar transistors.

- Field-effect transistors produce less electrical noise than bipolar transistors.

- Because of their unique mechanism, FETs are more resistant to the degrading effects of nuclear radiation.

- MOSFETs use less power than bipolar transistors, in some cases, about a thousand times less. This means fewer cooling requirements, using less space at less cost.

# JUNCTION FIELD EFFECT TRANSISTOR (JFET) CONSTRUCTION

Either N-channel or P-channel types can be used to describe the construction of a JFET. An N-channel JFET is used in the following explanation, but with the proper polarities, an identical explanation can be used to describe a P-channel JFET.

An N-channel JFET is constructed by initially diffusing the element phosphorus, in gaseous form, into a silicon wafer. This produces N-type material on each of the appropriately masked sections of the wafer. This diffusion creates the channel and provides the polarity designation to the JFET, in this case, an N-channel JFET.

The element boron, in gaseous form, is the second dopant diffused into the wafer to create P-type material in the appropriately masked areas. See Figure 69.

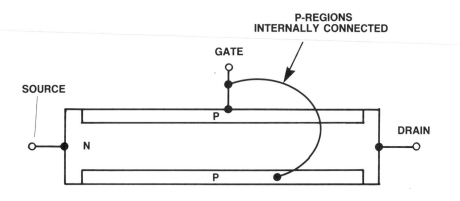

N-channel Junction FET Construction
Cross-section of a Single Chip
**Figure 69**

The FET chip is completed by bonding wires to the drain, source and gate contacts and packaged so that the elements of the FET are connected to external leads.

# JUNCTION FET MECHANISM

A simplified cross-sectional view and schematic circuit of the N-channel JFET circuit is shown in Figure 70.
- The drain-to-source path (channel) is the output section.
- The gate-to-source PN junction is the input section.
- The source is common to both input and output.
- The voltage supply, $V_{DD}$, is connected between drain and source through a load resistor, $R_L$, with positive polarity to the drain. (Note the use of double-subscript notation for the designation of the supply voltage.)

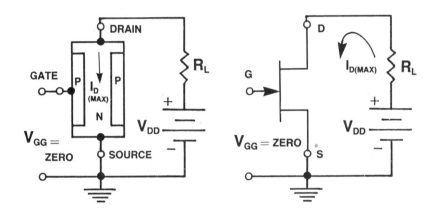

Initial Condition of JFET Cross-section and Circuit
**Figure 70**

With zero input voltage applied betwen gate and source, the channel is highly conductive (very low in resistance). The value of resistance is determined by the dimensions of the channel (W x L) and by the amount of dopant used to create the N-type material. Under this condition of zero input, the drain current, $I_D$, is at its highest value.

$$I_D = \frac{V_{DD}}{R_L}$$

Since the input (gate-to-source) PN junction is not forward biased, its resistance (the input resistance of the JFET) is essentially infinite. Since the channel resistance is at its lowest value, the output circuit is said to be in a state of *drain saturation*, while the input resistance is essentially infinite. Drain saturation allows maximum drain current, $I_{D(max)}$, to flow in the channel while still maintaining essentially infinite input resistance. This characteristic is specified on the data sheet as $I_{DSS}$.

The channel resistance could be reduced slightly further by applying a forward bias voltage (positive to the gate) to the gate-to-source PN junction. Forward biasing would also reduce the input resistance of the JFET to a low value. This, however, would negate the advantage of the FET compared to the bipolar transistor.

If a reverse bias voltage is applied between gate and source (negative to the gate), the resulting electric field would create a depletion area around each of the P-type materials between gate and channel. Since the depletion area is void of electrons and ions, it has extremely high resistance. The existence of the depletion areas reduces the width of the channel, resulting in an increase in channel resistance. An increase in channel resistance decreases the drain current, $I_D$. See Figure 71. The reverse bias on the input junction still maintains an infinite input resistance characteristic.

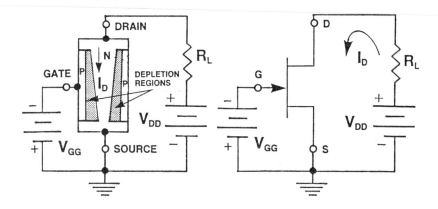

Reverse Biasing the JFET Input
**Figure 71**

As the reverse voltage between gate and source is increased, the depletion areas between gate and channel are increased, further increasing channel resistance. When the increasing reverse input voltage causes the depletion areas to touch, the channel is pinched off, increasing its resistance to infinity and cutting off the flow of drain current. See Figure 72.

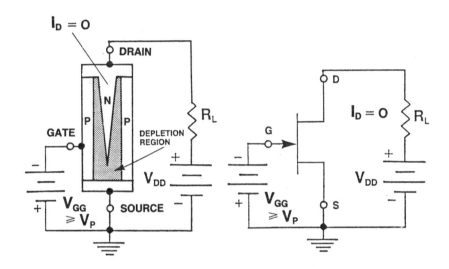

Cutting Off the Channel
**Figure 72**

At a specified value of $V_{DD}$, the level of reverse input voltage that causes the channel to cut off is called the *pinch-off voltage*, designated as $V_P$.

Because of the supply voltage, $V_{DD}$, a voltage gradient exists across the channel and causes the depletion areas to exhibit a nonuniform shape across the channel. See Figures 71 and 72. With different values of $V_{DD}$, a different gradient will exist and the channel will cut off at different values of $V_P$. The value of $V_P$ is therefore dependent on the value of $V_{DD}$ used.

A reduction in gate reverse voltage will cause the depletion areas to decrease, reducing channel resistance and resulting in increased drain current. When the input voltage is reduced to zero, drain saturation ($I_{DSS}$) is reached and drain current is at its maximum value. See Figure 70. The dynamic range of operation is between the limits of $I_{DSS}$ and channel cut-off.

# JUNCTION FET OPERATION AS A LINEAR AMPLIFIER

To operate the junction FET as a linear amplifier, an initial reverse bias must be applied between the gate and source to establish the quiescent output operating point. A second source of voltage ($V_{GG}$) can be used as the gate reverse bias around which an incoming signal can be varied. It is connected through resistor, $R_G$, to the gate terminal, providing a D.C. path between gate and source. See Figure 73.

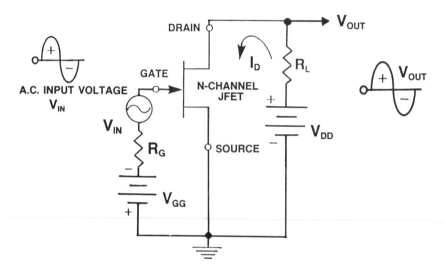

Junction FET Operating as a Linear Amplifier
**Figure 73**

Resistor, $R_G$, must be high enough in value compared with the output resistance of the preceding amplifier stage to avoid any loading effect on the preceding amplifier.

As a $\Delta V_{IN}$, the incoming signal from the preceding output circuit is applied to the JFET input, the gate-to-source voltage will vary by the amount of the input voltage variation, causing the drain (output) current to vary accordingly.

As the changing drain current, $\Delta I_D$, flows through the load resistor, $R_L$, an output voltage, $\Delta V_{OUT}$, is developed across $R_L$. As long as the input voltage remains within its dynamic limits ($V_{IN}$ = zero to pinchoff) the circuit will maintain linear amplification. Voltage gain is defined as a change in output voltage, $\Delta V_{OUT}$, divided by a change in input voltage, $\Delta V_{IN}$.

A more desirable circuit approach than that used in Figure 73, is to eliminate the separate voltage supply, $V_{GG}$, and to use a different way to achieve gate bias while providing some circuit stabilization.

This is accomplished by connecting a low value resistor and a capacitor in parallel between the source terminal and the reference to provide a self-biasing circuit. See Figure 74.

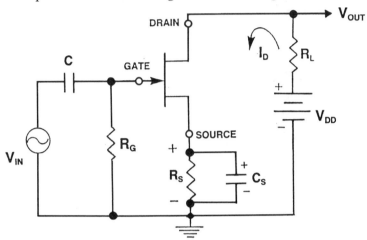

Stabilized Junction FET Amplifier With Self-Bias
**Figure 74**

- The drain current that flows through $R_S$ develops a voltage that provides the reverse bias between gate and source. Note the polarity of the voltage across $R_S$.
- Because of this voltage, the gate is more negative than the source and establishes the quiescent drain current point.
- As the incoming input voltage, $\Delta V_{IN}$, is applied to the gate terminal, the changing drain current, $\Delta I_D$, will produce a changing output voltage, $\Delta V_{OUT}$, across the load resistor, $R_L$.
- The function of the capacitor, $C_S$, is to filter out the variations across $R_S$ produced by the changing input voltage.

In addition, the self-biasing voltage developed across $R_S$ provides circuit stabilization despite changing temperature.
- A decrease in drain current because of an increase in temperature will result in a decrease in self-bias voltage.
- In turn, this bias voltage decrease will cause an increase in the quiescent drain current level opposing the change due to the temperature increase.

This action is called *negative feedback*, maintaining stabilized circuit operation regardless of changing temperature.

# JUNCTION FET OPERATION AS A DIGITAL SWITCH

## TURNED-ON STATE

With zero voltage applied to the input of a junction FET, the channel resistance is at its lowest value and is designated as $R_{DS(ON)}$. See Figure 75. If $R_{DS(ON)}$ is very small compared with $R_L$, it is considered negligible and assumed to be zero.

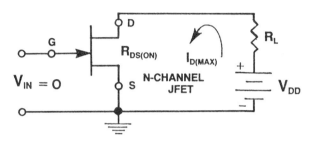

Junction FET - Turned-On State
**Figure 75**

When the JFET is ON, the output current, $I_D$, is at its maximum value and is equal to the supply voltage divided by the load resistor, $R_L$.

## TURNED-OFF STATE

- When a negative polarity voltage step is applied to the input terminals of the N-channel JFET, the channel resistance will increase in a step-function manner.
- If the magnitude of the input pulse is greater than the pinch-off voltage, $V_P$, the channel resistance will be essentially infinite and the JFET will turn OFF. See Figure 76.

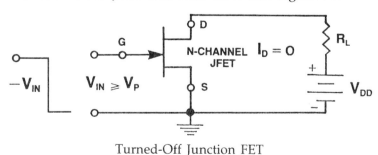

Turned-Off Junction FET
**Figure 76**

As the input returns to zero, the JFET reverts to its ON state.

# MOSFET — METAL-OXIDE SEMICONDUCTOR FIELD-EFFECT TRANSISTOR

There are two basic types of MOSFETs:
- Enhancement type
- Enhancement-Depletion type

## ENHANCEMENT TYPE MOSFET
(Used for switching applications)

- With zero input voltage applied to the gate, the MOSFET is OFF - the standby mode for the MOSFET.
- When an input digital pulse of the proper polarity is applied to the gate, the channel resistance will change from infinity to essentially zero. The channel becomes conductive, or is enhanced by the presence of the electric field created by the input voltage. When the MOSFET is turned ON, the channel resistance is at its lowest value and the output (drain) current is at its maximum value.
- Removal of the input pulse will cause the channel to revert to its original infinite resistance, turning the MOSFET OFF.

The action of the enhancement type MOSFET is analogous to a spring-loaded (momentary), normally OFF, mechanical switch.

## DEPLETION - ENHANCEMENT TYPE MOSFET
(Used for linear amplifier applications)

With zero input voltage applied to the gate, the channel resistance is at a specific moderate value and the drain current is at its quiescent level, waiting for an input voltage.

Depending on the polarity of the applied A.C. input voltage, the following will occur:
- The channel will be enhanced (its resistance will decrease), causing more drain current to flow; or
- The channel will be depleted (its resistance will increase), causing less drain current to flow.

To maintain linear amplification, the limits of the maximum input voltage range should be as follows:

- At the low end - the value of the pinch-off voltage, $V_P$.

- At the high end - the value of input voltage that will cause the MOSFET to saturate. Saturation is defined as that value of drain current where there is no further enhancement of the channel and, therefore, no further increase in drain current.

## ENHANCEMENT MOSFET CONSTRUCTION
## AND MECHANISM

Either N-channel or P-channel types can be used to describe the construction and mechanism of an enhancement type MOSFET. An N-channel type is used for this explanation. The same approach can be used with a P-channel type, with voltage polarities reversed.

- Starting with a silicon wafer and using boron as the dopant, the first diffusion creates the P-type area.
- The second diffusion creates N-type material in the areas indicated, using phosphorus as the dopant. Note that a section of the P-region lies between the two N-regions.

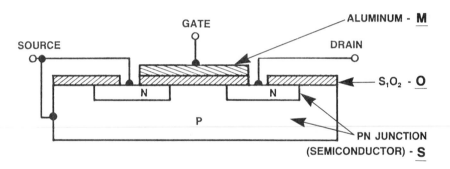

Construction of an Enhancement Type MOSFET
**Figure 77**

- An insulation layer of silicon dioxide is then grown on the surface of the N-regions and the P-region between them, with two windows in the silicon dioxide to allow access for terminals to be bonded to the N-regions.
- A layer of aluminum is then deposited on the silicon dioxide.
- Terminal leads are internally bonded to the appropriate parts of the chip and assigned designations as follows:
  — Gate terminal - to the aluminum section.
  — Drain and Source terminals - to the N-regions through the access windows in the silicon dioxide layer.

The composition of a MOSFET is structured as follows:
- The Gate is made of aluminum, a Metal.
- The silicon dioxide is an Oxide.
- The adjacent PN regions create a Semiconductor.

TURNED-OFF STATE (STANDBY MODE) (Figure 78a)
With zero input voltage applied to the gate and anode positive
voltage applied to the anode through a load resistor, the MOS-
FET is OFF - the standby mode for the MOSFET.
The resistance of the N-channel is essentially infinite because of
the reversed biased NP junction between the drain and the
adjacent P-region and no drain current will flow. With no drain
current flowing, the voltage between drain and source, $V_{OUT}$, is
equal to the voltage supply, $V_{DD}$.

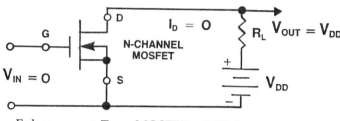

Enhancement Type MOSFET - OFF State
**Figure 78a**

TURNED-ON STATE (Figure 78b)
When a positive polarity input step voltage is applied to the gate
and a positive steady-state D.C. supply voltage is connected to
the drain through a load resistor, $R_L$, the MOSFET will turn ON.

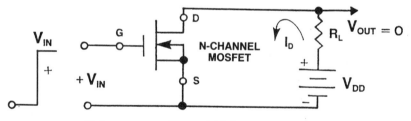

Enhancement Type MOSFET - ON State
**Figure 78b**

The sufficiently high input step voltage creates an electric field
across the silicon dioxide layer existing between gate and channel, causing the P-region between the two N-regions to totally
convert from P-type to N-type material. The channel is enhanced
(becomes conductive), closing the drain circuit and allowing
saturated drain current to flow. With the MOSFET in its ON
state, the voltage between drain and source, $V_{OUT}$, is equal to
zero.

The change in output voltage, $V_{OUT}$, can be used to switch other
stages involved in a complex switching sequence.

It should be noted, that a MOSFET, when in its conducting state, has its input section (gate-to-channel) made of an aluminum conducting area separated from a conductive N-channel by a layer of nonconducting silicon dioxide. Conceptually, a capacitor is made up of two conducting areas sandwiching a nonconductive area between them. Therefore, the input section of the MOSFET functionally contains a built-in coupling capacitor.

With this input circuit structure, the input resistance of a MOSFET is essentially infinite and there is no loading of preceding circuitry. The input resistance is so high, however, that the device is extremely sensitive to the presence of any electrostatic energy in its proximity. If this high-voltage, low-current energy is discharged through the gate circuit it can cause immediate destruction of the MOSFET by puncturing the very thin silicon dioxide layer in the input section.

Many steps have been taken to protect the device from this destructive force, such as protective packaging, care in handling, proper grounding of assembly lines and wearing of clothing that would not generate electrostatic energy. The MOSFET can be more readily protected by diffusing a PN junction between the gate and source, thereby providing a protective built-in zener diode. See Figure 79.

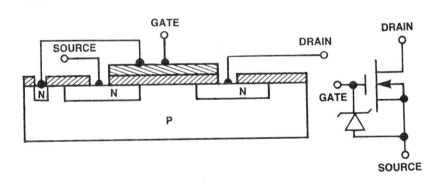

Protecting the MOSFET with a Built-in Zener Diode
**Figure 79**

If any electrostatic discharge (ESD) occurs, the zener diode will clamp that voltage at some safe level before it punctures the silicon dioxide layer, destroying the device. The zener diode will clamp at any voltage greater than the magnitude of the input voltage. The only effect on the circuit will be an insignificant decrease in input resistance.

# ENHANCEMENT-DEPLETION MOSFET

## CONSTRUCTION AND MECHANISM

Either N-channel or P-channel types can be used to describe the construction and mechanism of an enhancement-depletion type MOSFET. The N-channel type is used for this explanation. The same approach can be used with a P-channel type, with voltage polarities reversed.

The construction of the enhancement-depletion type MOSFET is essentially the same as the enhancement type. The difference is that the N-regions that previously were totally blocked by the P-region at zero input voltage, are now only partially blocked by the P-region. See Figure 80.

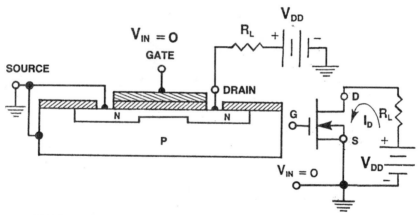

Enhancement-Depletion MOSFET with Zero Input Voltage
**Figure 80**

With the channel having a specific resistance at zero input voltage, a quiescent drain current will be established for the MOSFET to operate as a linear amplifier. This quiescent drain current develops a quiescent voltage across load resistor, $R_L$.

Referring to Figure 81, when an A.C. input voltage is applied between the gate and source, the changing electric field produced by the changing input voltage will modify the conductivity of the channel as follows:
- When the input A.C. voltage is positive, the channel is enhanced, its resistance decreases, more drain current flows and the voltage across the load, $V_{OUT}$, increases.

- When the input A.C. voltage is negative, the channel is depleted, its resistance increases, less drain current flows and the voltage across the load, $V_{OUT}$, decreases.

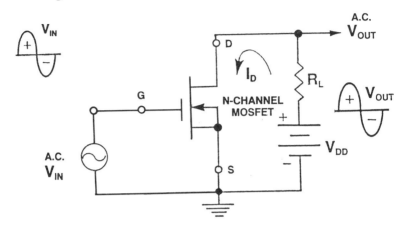

Enhancement-Depletion MOSFET with Input Voltage Applied
**Figure 81**

The change in drain current produced by a change in input voltage is called *common source forward transconductance* and is designated as $g_{fs}$ and measured in micromhos. A mho is a unit of conductance (the reciprocal of ohm, the unit of resistance).

$$\text{Transconductance} = g_{fs} = \frac{\Delta I_D}{\Delta V_{IN}}$$

Voltage gain in the amplifier is defined as the changing output voltage, $\Delta V_{OUT}$, divided by the changing input voltage, $\Delta V_{IN}$.

$$\text{Voltage Gain} = \frac{\Delta V_{OUT}}{\Delta V_{IN}}$$

In this linear amplifier, as in all previous examples of linear amplification, a sine wave is used as the input signal. This does not imply that only a sine wave can be used for linear amplification. All varieties of A.C. can be amplified in linear form and include square waves, saw-tooth waves, and other non-sinusoidal waveshapes (voice frequencies and music).

# JUNCTION FIELD-EFFECT TRANSISTOR SPECIFICATIONS

To assist in the selection of a proper device for a particular application, data sheet information lists the following:
- Device polarity - N-channel or P-channel.
- Material and manufacturing process.
- Physical outline and dimensions.
- Electrical features and general and/or unique applications.

The major specifications are classified under the headings of *absolute maximum ratings* and *electrical characteristics*.

POLARITY

N-channel JFET - positive voltage on the drain
                 negative voltage on the gate

P-channel JFET - negative voltage on the drain
                 positive voltage on the gate

Selection of JFET polarity is based on:
- The polarity of the supply voltage with respect to ground.
- The polarity of the D.C. signal to be amplified.
- The polarity of the switching voltage used to turn the JFET OFF when it is being used as a digital switch.

Absolute maximum ratings are the electrical, mechanical and thermal limits of the device assigned by the manufacturer. If these ratings are exceeded, permanent damage to the device may result. These values are used as limiting values and do not specify the actual operating conditions of the transistor.

Electrical characteristics are the measurable electrical properties of the device inherent in its design and are listed at a specific operating temperature. The characteristics describe the features of the device under conditions which approximate its actual operation.

## ABSOLUTE MAXIMUM RATINGS - JFET

GATE-TO-DRAIN VOLTAGE OR GATE-TO-SOURCE VOLTAGE - $V_{GD}$ OR $V_{GS}$

This is the maximum reverse voltage between gate and drain or gate and source before breakdown occurs. Effectively, it is the peak inverse voltage rating (PIV) of the gate to channel diode. Typically, this value ranges between 25 and 50 volts.

Almost all JFETs have a symmetrical geometry channel and the source and drain leads may be interchanged as long as the correct voltage polarity is used for the designated terminal.

DRAIN CURRENT - $I_{Dmax}$
Maximum allowable D.C. drain current. This value will depend on the size of the chip and package.
- Small-signal devices - from 10 to 50 milliamperes.
- Power devices - from 500 to 900 milliamperes.

GATE CURRENT - $I_{G(max)}$
Maximum allowable D.C. gate current. Although this parameter is generally listed on a data sheet, the gate circuit is rarely, if ever, set into a forward biased state.

POWER DISSIPATION - $P_D$
Maximum allowable power dissipation at a case temperature of 25°C. This 25°C rating must be derated as the transistor case temperature increases.

> *Power derating factor* is defined as the decrease in power capability as a function of increasing case temperature and is measured in watts per °C. The maximum power rating is generally listed at 25°C case temperature with no further allowable increase in power at lower temperatures.

Refer to CHAPTER FIVE — THYRISTORS for a discussion of heat sinking considerations for power devices.

OPERATING AND STORAGE TEMPERATURE RANGE
- For military and aerospace operation: -55°C to +125°C. Storage temperature: -55°C to +150°C.

- For commercial, industrial and consumer applications, there are no standard limits. The temperature range is typically from -10°C to +100°C.

The packages used for JFETs are the same as those used for bipolar transistors and they are available for commercial and military/space applications. The commercial types generally use a plastic (epoxy) package, with the military and aerospace types enclosed in a hermetically-sealed metal package.

Small-signal devices are generally used with no heat sinks. Power devices are usually mounted to appropriate heat sinks.

# ELECTRICAL CHARACTERISTICS - JFET

Electrical specifications are listed at a temperature of 25°C, unless otherwise noted. In addition to typical values, minimum and maximum values are given where applicable.

## GATE REVERSE CURRENT - $I_{GSS}$
The gate-to-source leakage current with the drain shorted to the source at a specified $V_{GS}$. In many cases, values for $I_{GSS}$ are given at elevated temperatures. $I_{GSS}$ values approximately double for every 10°C rise in temperature.
- Small-signal JFETs - typically from 0.1 to 2 nanoamperes.
- Power JFETs - typically from 2 to 5 nanoamperes.

## DRAIN CUTOFF CURRENT - $I_{D(OFF)}$
The maximum drain leakage current under a condition of cutoff (gate voltage equal to or greater than the pinch-off value) at a specified $V_{DD}$ and $V_{GS}$. $I_{D(OFF)}$ indicates the quality of the device as a turned-off switch. Maximum values typically range from 0.5 to 2 nanoamperes for small-signal JFETs. Power JFETs range slightly higher.

## DRAIN CURRENT AT ZERO GATE VOLTAGE - $I_{DSS}$
This is the value of drain current with the gate shorted to the source at a specified $V_{DD}$. It is also called the drain saturation current and indicates the maximum resistance of the device in its ON state when used as a switch.

## GATE-TO-SOURCE PINCH-OFF VOLTAGE - $V_P$
This is the gate-to-source reverse bias voltage required to reduce drain current to essentially zero at a specified $V_{DD}$ and $I_{D(OFF)}$. Typical values range from 0.5 to 10 volts.

## COMMON SOURCE FORWARD TRANSCONDUCTANCE - $g_{fs}$
This value is an indication of the voltage amplification capability of the device. It is the change in the drain current, $\Delta I_D$, produced by the change in the input, or gate voltage, $\Delta V_{IN}$. The changing drain current flowing through the load resistor develops the output voltage. This parameter is measured at a constant value of $V_{DD}$ and at a specified $V_{GS}$. Typical values range from 3000 to 10,000 micromhos.

## DRAIN-TO-SOURCE ON RESISTANCE - $R_{DS(ON)}$ or $R_{ON}$
This is the resistance value of the channel with zero gate voltage ($I_{DSS}$) and at a specified $V_{DD}$. Typical values range from 1 to 100 ohms.

## GATE-TO-SOURCE INPUT CAPACITANCE - $C_{iss}$

This value is measured with the source connected to the drain at specified values of $V_{GS}$ and frequency. The value of $C_{iss}$ depends largely on the parallel combinations of capacitance produced by the geometry of the chip and the structure of the package. The value of $C_{iss}$ is also dependent on the value of $V_{GS}$. The higher the $V_{GS}$, the lower the capacitance. Depending on the device and specified operating conditions, typical values range from 1 to 40 picofarads.

For small-signal operation, quiescent $V_{GS}$ is sometimes set near pinch-off voltage to minimize drain current (and power) and also to minimize $C_{iss}$ which maximizes frequency response.

## DRAIN-TO-SOURCE OUTPUT CAPACITANCE - $C_{oss}$

This value is given with the source connected to the gate, or $V_{GS}$ equal to zero at a specified $V_{DD}$ and frequency. The higher the value of $V_{DD}$, the higher the output capacitance. In addition, $C_{oss}$ depends to a large extent on the geometry of the chip and package. Typical values for $C_{oss}$ range from 0.2 picofarads to 1.0 picofarads.

## TURN-OFF TIME - $T_{OFF}$

When the input step voltage reaches or exceeds pinch-off voltage (voltage for the N-channel JFET and positive voltage for the P-channel JFET), there is a finite interval of time needed for the JFET to turn OFF. This is the time between the start of the reverse input step voltage and the time when the drain current has decreased to 10% of its initial value.

## TURN-ON TIME - $T_{ON}$

When the JFET is turned ON by removing the reverse step voltage that had pinched off the device, there is a finite time for the channel to respond and become conductive. This time is called the turn-on time and includes both *delay time* and *rise time*.

- Delay time ($t_d$) is the interval between the start of the input voltage return to zero and the time when the drain current reaches 10% of its final value.

- Rise time ($t_r$) is the interval between the 10% and 90% values of the final drain current.

If turn-on and turn-off times are 50 nanoseconds or less, the device is classified as a computer switch. If these times are longer, it is specified as an amplifier or as a general-purpose switch.

# MOSFET SPECIFICATIONS

To assist in the selection of a proper device for a particular application, the manufacturer's data sheet information lists the following:

- Device polarity - N-channel or P-channel.
- Type - enhancement or enhancement-depletion.
- Gate protection (zener clamp), if applicable.
- Material and manufacturing process.
- Physical outline and dimensions.
- Significant electrical features and their applications.

The major specifications are classified under the headings of *absolute maximum ratings* and *electrical characteristics*.

POLARITY
N-channel MOSFET - positive voltage on the drain.
Enhancement type:    positive voltage on the gate to enhance the channel.

P-channel MOSFET - negative voltage on the drain.
Enhancement type:    negative voltage on the gate to enhance the channel.

Selection of MOSFET polarity is based on:
- The polarity of the supply voltage with respect to source.
- The polarity of the D.C. signal to be amplified.
- The polarity of the switching voltage used to turn the MOSFET OFF when it is being used as a digital switch.

Absolute maximum ratings are the electrical, mechanical and thermal limits of the device assigned by the manufacturer. If these ratings are exceeded, permanent damage to the device may result. These are used as limiting values and do not specify the actual operating conditions of the transistor.

Electrical characteristics are measurable electrical properties of the device inherent in its design and are listed at a specific operating temperature. The characteristics describe the features of the device under conditions which approximate its actual operating mode.

Electrical specifications are listed at a temperature of 25°C, unless otherwise noted. In addition to typical values, minimum and maximum values are given where applicable.

## ABSOLUTE MAXIMUM RATINGS - MOSFET

### DRAIN-TO-SOURCE VOLTAGE - $V_{DS}$
Maximum allowable voltage across the channel. Indicates the maximum supply voltage, $V_{DD}$, that can be used. Ranges between 30 and 250 volts D.C. (New devices extend to 1000 volts.)

### GATE-TO-DRAIN VOLTAGE OR GATE-TO-SOURCE VOLTAGE - $V_{GD}$ OR $V_{GS}$
- For devices that are not gate-protected, this is the maximum reverse voltage between gate and drain or gate and source before the silicon dioxide layer is punctured. This value is about 50 to 75 volts D.C.
- For gate-protected devices, this is the reverse breakdown voltage of the protective zener and is about 30 volts.

### DRAIN CURRENT - $I_{D(max)}$
Maximum allowable D.C. drain current. This value depends on the size of the chip and package.
- Small-signal devices - from 10 to 50 milliamperes.
- Power devices - from 500 to 900 milliamperes.
  (New power devices range from 2.5 to 50 amperes.)

### GATE CURRENT - $I_{G(max)}$ FOR INTEGRATED ZENER DIODE CLAMP
Maximum allowable D.C. gate current. This is the maximum current of the built-in protective zener diode. When no protective zener diode is used, this rating is omitted.

### POWER DISSIPATION - $P_D$
Maximum allowable power dissipation at a case temperature of 25°C. This rating must be derated as the temperature of the case increases. Power MOSFETs range from 5 to 250 watts.

Refer to CHAPTER FIVE — THYRISTORS for a discussion of heat sinking considerations for power devices.

### OPERATING AND STORAGE TEMPERATURE RANGE
For military and aerospace operation: -55°C to +125°C.
  Storage temperature range: -55°C to +150°C.
For commercial, industrial and consumer applications, there are no defined limits. The temperature range is determined by the manufacturer and typically extends from -10°C to +100°C.

The packages used for MOSFETs are the same as those used for bipolar transistors. They are available for both commercial and military/space applications. The commercial types generally use an epoxy package, with the military/space types enclosed in a hermetically-sealed metal package.

## ELECTRICAL CHARACTERISTICS - MOSFET

### DRAIN-TO-SOURCE ON RESISTANCE - $R_{DS(ON)}$ or $R_{ON}$ - (ENHANCEMENT)
The resistance value of the enhanced channel at a specified $V_{GS}$, $I_D$ and the body (substrate) connected to the source or to a specified body-to-source voltage.
● Small-signal devices values range from 10 to 1000 ohms.
● Power MOSFET values range from .01 ohms to 0.1 ohms.
Since the enhancement-depletion type is only used as a linear amplifier, this characteristic is not listed for this type.

### GATE THRESHOLD VOLTAGE - $V_{GS(TH)}$
The value of gate-to source voltage that must be exceeded to set the channel to its enhanced (conductive) state. It is measured at a specified value of $I_D$, with the body connected to the source. Maximum values are from 4.0 to 6.0 volts.

### DRAIN CUTOFF CURRENT - $I_{D(OFF)}$
This parameter specifies the maximum drain leakage current with zero gate-to-source voltage at a specified $V_{DS}$, with the body connected to the source. It indicates the quality of the device as a turned-off switch. The maximum values range from 0.1 to 10 nanoamperes.

### GATE-TO-BODY LEAKAGE CURRENT - $I_{GSS}$
This parameter specifies leakage current between gate and body at a specified $V_{GS}$ with the drain and body connected to the source. In many cases, values for $I_{GSS}$ are specified at elevated temperatures. $I_{GSS}$ approximately doubles for every 10°C rise in temperature.
● Small-signal: Values range from 0.001 to 1.0 nanoampere.
● Power: Values range from 1 to 5 nanoamperes.

### DRAIN CURRENT IN ENHANCED STATE - $I_{D(ON)}$
This is the value of drain current with the channel in a saturated enhanced state at a specified value of $V_{DS}$ and $V_{GS}$ and the body connected to the source. It indicates the quality of the device in its ON state when used as a switch.

## COMMON SOURCE TRANSCONDUCTANCE - $g_{fs}$ - (ENHANCEMENT-DEPLETION)

This value is the change in drain current, $\Delta I_D$, caused by the change in input voltage, $\Delta V_{IN}$, at a specified value of $V_{DS}$, $V_{GS}$ and frequency, with the body (substrate) connected to the source. Typical values range from 1000 to 4000 micromhos.

## GATE-TO-SOURCE CAPACITANCE - $C_{GS}$

This value is measured with the source connected to the body at specified values of $V_{DS}$, $V_{GS}$ and frequency. $C_{GS}$ depends largely on the geometry of the chip and package in addition to the values of $V_{DS}$ and $V_{GS}$. As these voltage values are increased, $C_{GS}$ is decreased. Typical values range from 0.1 to 0.5 picofarads.

## GATE-TO-DRAIN CAPACITANCE - $C_{GD}$

This value is measured with the drain connected to the body at specified values of $V_{DS}$, $V_{GD}$ and frequency. $C_{GD}$ depends largely on the geometry of the chip and package in addition to the values of $V_{DS}$ and $V_{GD}$. As these voltage values are increased, $C_{GD}$ is decreased. Typical values range from 0.1 to 0.5 picofarads.

## SOURCE-TO-BODY or DRAIN-TO-BODY CAPACITANCE - $C_{SB}$ or $C_{DB}$

This is measured with the gate shorted to the body, at a specified frequency and with $V_{GB} = V_{DB}$ at a specified voltage. Typical values range between 1.0 and 2.0 picofarads.

## DRAIN-TO-SOURCE CAPACITANCE - $C_{DS}$

This value is at a specified $V_{DS}$, gate-to-body voltage, $V_{GB}$, drain-to-body voltage, $V_{DB}$, source-to-body voltage, $V_{SB}$ and frequency. $C_{DS}$ depends to a large extent on the geometry of the chip and package in addition to the value of these voltages. As the voltage increases, $C_{DS}$ decreases. Typical values for $C_{DS}$ range from 0.2 picofarads to 1.0 picofarads.

## TURN-ON TIME - $T_{ON}$

When the channel is enhanced by applying a step input voltage (positive for the N-channel and negative for the P-channel), a finite time exists for the channel to respond and become conductive. This is called turn-on time and is characterized in the same manner as specified for JFETs.

## TURN-OFF TIME - $T_{OFF}$

When the input returns to zero, there is a finite time interval for the MOSFET to turn OFF. This is called turn-off time, characterized in the same manner as specified for JFETs.

# CMOSFET (COMPLEMENTARY MOSFET) TECHNOLOGY

When N-channel and P-channel enhancement type MOSFETs are connected to form a single digital switch, the combination is called Complementary MOSFET or CMOSFET. See Figure 82.

- The positive terminal of supply voltage, $V_{DD}$, is connected to the source terminal of the P-channel MOSFET.
- The gates of the complementary N-channel and P-channel MOSFETs are connected internally to an external input terminal.
- The source terminal of the N-channel MOSFET is connected to the negative terminal of the voltage supply, $V_{DD}$. Although the voltage supply is applied to the source of the CMOSFET, because of custom, it is still referred to as $V_{DD}$.
- The drains of the complementary N-channel and P-channel MOSFETs are internally connected to form a common drain terminal.

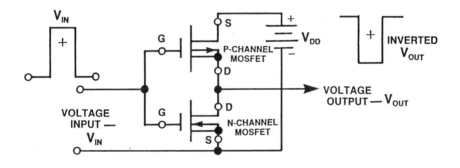

CMOSFET (Complementary MOSFET)
**Figure 82**

## CMOSFET OPERATION

$V_{IN}$ AT ITS HIGH LEVEL ($V_{IN} = V_{DD}$)
- When a positive step voltage equal to $V_{DD}$ is applied between gate and source, the channel of the N-channel MOSFET is enhanced, causing it to turn ON.

- At the same time, the existence of a positive input voltage produces a condition of zero voltage between the P-channel gate and its source. Since it is an enhancement (normally OFF) type, the channel resistance of the P-channel MOSFET is extremely high and acts as the high resistance load for the turned-on N-channel MOSFET.

- With the channel resistance of the turned-on N-channel MOS-FET having essentially zero resistance, its drain and source are at the same point electrically (zero voltage with respect to the reference. The output is therefore at zero voltage.

$V_{IN}$ AT ITS LOW LEVEL ($V_{IN}$ = ZERO)
- With the input digital pulse at zero level, the source of the P-channel MOSFET is more positive than its gate (or the gate is more negative than its source), enhancing the channel of the P-channel MOSFET, causing it to turn ON.

- At the same time, since the input voltage at its zero level, the voltage between gate and source of the N-channel MOSFET is also equal to zero, turning the N-channel MOSFET OFF. The channel resistance of the N-channel MOSFET is then extremely high and acts as the high resistance load for the turned-on P-channel MOSFET.

- With the channel resistance of the turned-on P-channel MOS-FET at essentially zero resistance, its drain and source are at the same point electrically ($V_{DD}$ with respect to the reference). The output voltage is therefore equal to $V_{DD}$ with respect to the reference.

It should be noted that the CMOSFET switch acts as an inverting switching circuit. It responds to the input digital pulse by producing the opposite logic at its output:
- When the input is positive, the output is zero.
- When the input is zero, the output is positive.

It should also be noted that there is no external resistor in the output circuit connected between the supply voltage and the output terminal. In CMOSFET circuitry, there is no need for an external load resistor since the two complementary N-channel and P-channel MOSFETs act alternately as the load resistance.
- When the input voltage is positive, the P-channel MOSFET is the load resistance for the turned-on N-channel MOSFET.
- When the input is zero, the N-channel MOSFET is the load resistance for the turned-on P-channel MOSFET.

This feature of CMOSFET technology makes it applicable for high-density, low-power operation since it provides for:
- Simplified circuitry with a resulting reduction of parts.
- Extremely low power consumption because the high resistance of the channels allows very little output current to flow.

# FIELD-EFFECT TRANSISTOR APPLICATIONS

It is obvious that the high-resistance input characteristics of field-effect transistors make these devices very attractive for use in multi-stage voltage amplifier circuitry. No loading effects are exhibited and circuit design is simplified to a great extent. Junction FETs enhancement-depletion type MOSFETs are used for this group of applications.

The enhancement type MOSFETs are normally-off switches and are used in digital logic switching applications. In combining both N-channel and P-channel MOSFETs in a single CMOSFET, an effective built-in load resistor is provided. This technique requires fewer parts and offers lower power dissipation, however, small-signal CMOS circuits are about ten times slower than small-signal bipolar transistor digital logic circuits. In contrast, power bipolar transistor circuits are much slower than similar power MOSFET circuitry. See CHAPTER NINE — TRENDS FOR DISCRETE SEMICONDUCTORS.

## ANALOG SWITCH

The enhancement type MOSFET is also used as an *analog switch* - a digitally-controlled switch that is turned ON by a digital pulse and turned OFF by the removal of the pulse. Once turned ON, the channel provides the conducting path for low-level variable or A.C. voltages. These voltages are generally referred to as analog voltages. See Figure 83.

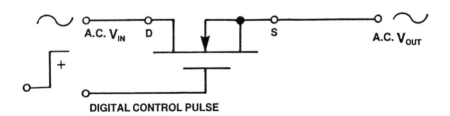

Analog Switch
**Figure 83**

In the past, analog switching was handled by electromechanical relays, with switching speeds in milliseconds. In contrast, FETS switch in nanoseconds and are smaller, lighter, have no contact bounce, are less expensive and are much more reliable.

## VIDEO AMPLIFIER

The field-effect transistor is particularly applicable for video amplifier use. Its normal frequency response is in excess of 250 Megahertz with some devices capable of operating as high as 450 Megahertz. The input resistance of a FET operating at these high frequencies is generally in the 10 Megohm range with input capacities reduced to less than 1 picofarad by standard circuit techniques. Both JFET and MOSFET video amplifiers are often used in communications and pulse amplifying circuits operating at frequencies of 100 Megahertz and above.

## VOLTAGE CONTROLLED RESISTOR

If the proper reverse bias voltage is applied to the gate terminal of a JFET, the channel resistance is set to some specific value depending on the value of the quiescent $V_{GS}$.

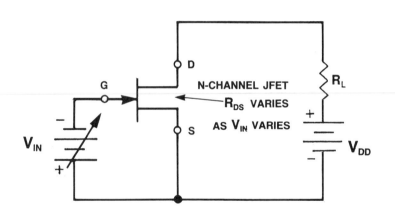

FET Used as a Voltage-Controlled Resistor
**Figure 84**

If the input voltage, $V_{IN}$, is varied above and below the quiescent $V_{GS}$, the channel resistance will vary accordingly.

- As $V_{GS}$ decreases, the channel resistance decreases.
- As $V_{GS}$ increases, the channel resistance increases.

The quiescent voltage point and the range of input voltage variation must be carefully selected to achieve a linear change in channel resistance as a function of changing input voltage.

## VOLTAGE-SENSOR FOR TIMING CIRCUIT

Accurate, low-cost voltage-sensing in timing circuits can use the high-resistance input section of a JFET or MOSFET to sense a voltage applied to a resistor-capacitor (RC) network. The FET acts to monitor the charging and discharging voltage without loading the RC network. The output resistance, $R_L$, must be relatively low to efficiently drive succeeding circuitry.

The heart of most voltage sensing and timing circuits is the resistor-capacitor network. Time periods are determined by the charge and discharge of a capacitor through a resistor, with both the charge and discharge following a classical exponential pattern. The applied voltage is monitored by a FET voltage sensor which has a high resistance input circuit with very low capacitance. These characteristics will prevent any detrimental influence on the RC time constant. See Figure 85.

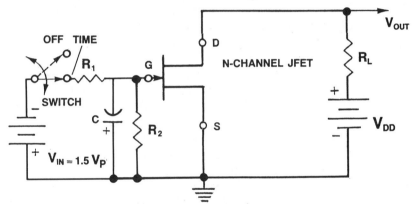

Voltage Sensor for Timing Circuit
**Figure 85**

- With the capacitor, C, fully discharged, and switch, $S_1$, set in the OFF position, $V_{GS}$ = zero and the channel resistance is essentially zero. Under this condition, the output voltage, $V_{OUT} = V_{DS}$, is at zero.

- When the switch is set to the TIME position, and the applied voltage, $V_{IN}$, equal to about 1.5 x the pinch-off voltage, $V_P$, of the FET, the capacitor will begin to charge. A single time constant is equal to $R_1$ x C, with $R_1$ in ohms and C in farads.

- When the capacitor is sufficiently charged so that $V_{GS} = V_P$, the FET will cut off. With drain current no longer flowing, the output voltage, $V_{OUT} = V_{DS} = V_{DD}$.

- With the switch reset to its OFF state, the capacitor will discharge through $R_2$, reducing the reverse bias between gate and source. As the reverse bias decreases, channel resistance decreases, causing drain current to increase until the output voltage, $V_{OUT}$, is essentially zero. The total discharge time is equal to 6 x ($R_2$ x C) with $R_2$ in ohms and C in farads.

By effectively isolating the RC network in the FET gate circuit from succeeding circuits that might otherwise present variable loading, an accurate RC time constant can be maintained.

## CONSTANT-CURRENT SOURCE

The JFET may be used as a constant-current source, maintaining a selected value of current, $I_D$, over fairly wide variations of supply voltage, $V_{DD}$, and load resistance, $R_L$.

When the drain-to-source voltage, $V_{DS}$, is significantly greater than the pinch-off voltage, $V_P$, the JFET may be reverse biased to operate as a constant-current source at any current below its saturation current, $I_{DSS}$. The series resistor, $R_S$, in the source circuit establishes the proper value of reverse bias to set the desired drain current, $I_D$. The voltage, $V_S$, developed across the series resistor, $R_S$, provides the reverse bias between gate and source. $V_S = - V_{GS}$.

- An increase in supply voltage will tend to increase the drain current. This will cause $V_{GS}$ to increase, maintaining drain current at its original value.
- A decrease in supply voltage will tend to decrease the drain current. This will cause $V_{GS}$ to decrease, maintaining drain current at its original value.
- Variations in load resistance will have the same effect and a constant drain current will be maintained.

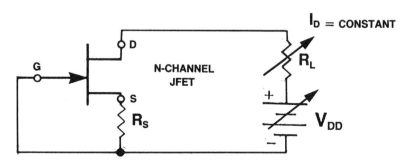

FET used as Constant-Current Source
**Figure 86**

# REINFORCEMENT EXERCISES

Answer TRUE or FALSE to the following statements:

1. One of the major features of a field-effect transistor is its extremely low input resistance characteristic which provides no loading of preceding circuitry.

2. Field effect transistors are used as voltage amplifiers, digital switches, analog switches and voltage-controlled resistors.

3. There are two basic types of FETs - the junction FET (JFET) and metal-oxide semiconductor FET (MOSFET), each having very high input resistance, but each having a different operating mechanism.

4. The JFET has a built-in diode at its input section which needs forward biasing to make it operate properly as an amplifier. In this state, its input resistance is very high.

5. The MOSFET has a built-in coupling capacitor at its input section, therefore, no external coupling capacitor is used to transfer an output voltage from a preceding circuit to the input section of the MOSFET.

6. The JFET can be used either as part of a voltage amplifier circuit, as a voltage-controlled resistor, or as part of a digitally-controlled switching circuit. Its main use, however, is in switching applications.

7. When used as a switch, the JFET is similar in action to a momentary type, normally-closed, single-pole, single-throw mechanical switch. With zero input voltage, drain current is flowing and power is being dissipated in the circuit.

8. The enhancement-depletion type MOSFET is used as a small-signal linear voltage amplifier having a very high input resistance characteristic and a built-in coupling capacitor at its input circuit. An incoming signal will either enhance or deplete its channel, varying drain current.

9. The action of an enhancement type MOSFET is similar to a momentary type, normally-open, single-pole, single-throw mechanical switch. During its standby state, no current is in the output section and no power is being dissipated.

10. The enhancement type MOSFET is a digital switch used in digital logic circuitry or as an analog switch to turn low-level A.C. and other low-level varying voltages ON and OFF.

11. A complementary metal-oxide semiconductor FET (CMOS-FET) is a component that combines an interconnected single N-channel MOSFET and a single P-channel MOSFET in one device.

12. With CMOSFET technology, the need for an external load resistor is eliminated since the two complementary N-channel and P-channel MOSFETs act alternately as the load resistance when a digital pulse is applied to the input.

13. A CMOSFET digital switch does not invert the logic of the input digital pulse. With zero voltage at the input, the output voltage is zero.

14. CMOSFET circuitry uses more power than either N-channel or P-channel circuits.

15. Because of its high input resistance characteristics, the field-effect transistor, either JFET or MOSFET, is appropriate for use in voltage sensing and timing circuits.

Answers to the reinforcement exercises are on page 194.

CHAPTER
EIGHT

# SEMICONDUCTOR RELIABILITY CONSIDERATIONS

MILITARY SPECIFICATION
MIL-S-19500

RELIABILITY LEVELS

MILITARY STANDARD MIL-STD-750

QUALIFICATION TO MIL-S-19500

THE ELEMENTS OF MIL-S-19500

QUALITY CONTROL PROGRAM

CAUSES OF SEMICONDUCTOR
FAILURE

GENERAL COMMENTS ON
HIGH-RELIABILITY SEMICONDUC-
TORS

# SEMICONDUCTOR RELIABILITY CONSIDERATIONS

## INTRODUCTION

The concept of reliability is based upon the ability of a component and the system in which it is assembled to perform as intended, regardless of its environment. Throughout its intended life, the component must conform to the electrical, mechanical and thermal requirements of its data sheet or the appropriate specification control drawing.

With the multitude of requirements that exist in the complex electronic systems used in today's military and aerospace equipment, the reliability of semiconductor components—the integral parts of these systems—must be considered a necessary subject for examination.

Reliability requirements for all components are addressed to:
- The minimization, and if possible, the elimination of the actual or potential failure mechanism in the composition and structure of the device.
- Maintaining the electrical stability of components throughout their operating life.
- Resisting the contaminating and destructive effects of any environment encountered in military/aerospace applications.

Potential reliability problems are inherent in:
- The fundamental design of the component.
- The manufacturing processes and materials used.
- Human error and/or machine malfunction in fabrication, assembly, pre-conditioning, testing and packaging.

In comparing passive and semiconductor component technologies, passive components are the older and more mature of the two and, to a great extent, the reliability problems of passive devices have been resolved. Semiconductor reliability, however, is of great concern for circuit designers, reliability specialists and all other personnel in the companies involved in the design and production of military and aerospace systems. It is these systems that demand the highest level of attention to component quality and longevity.

To establish an organized source of electronic components for use in military equipment, a centralized federal government agency was needed to achieve and maintain the following:

- A pre-determined level of reliability (quality and longevity) for electronic components and systems.
- Standardization of pre-conditioning techniques and test procedures in the production of components.
- Electrical and mechanical interchangeability of components.
- Standard inspection procedures and quality control methods.
- Documentation to assure that specified requirements are met.
- Aid in procurement of components.

In 1962, the Defense Electronics Supply Center (DESC) was created to coordinate the management, procurement and storage of all electronic components and equipment for military applications. It is an arm of the U.S. Government Department of Defense (DOD) and employs one of the highest concentrations of technically skilled personnel in the country. It is staffed with engineers, equipment and quality-control specialists and technicians to assure that the electronic requirements of the military be met.

Eventually, DESC became the sole government agency accountable for all electronic component procurement for the National Aeronautical and Space Administration (NASA), the Federal Aviation Agency (FAA) and the Weather Bureau, in addition to being an agency in the Department of Defense. DESC is responsible for initiating the specifications, standards and test procedures for all electronic components and equipment used for military and aerospace applications.

In cooperation with DESC, the Defense Contract Administration Service (DCAS), also operating under the jurisdiction of the Department of Defense, is concerned with the monitoring of the suppliers of those products and can impose their own controls. DCAS is responsible for the approval certification of the manufacture and quality status of standard military electronic products and for all Government Source Inspection (GSI).

Only the Department of Defense has the option to impose Government Source Inspection on any incoming component or manufacturing facility and can levy the requirement of GSI when necessary. A resident (or visiting) Government Source Inspector establishes the conditions of test (incoming inspection) and has complete responsibility to accept or reject any lot.

# MILITARY SPECIFICATION MIL-S-19500

MIL-S-19500 is the military specification governing discrete semiconductors and is defined by the conditions of the military guidelines standard, MIL-STD-750, with an appropriate suffix letter to denote the latest revision.

# RELIABILITY LEVELS

MIL-S-19500 establishes the general requirements for discrete semiconductors. Other detailed requirements and characteristics of a particular part are listed in an additional detail specification, called a *slash sheet*. Additional requirements for product assurances are based on differences in reliability levels for various types of systems. Each level is designated by a prefix added to the standard JEDEC part number that becomes part of that device identification. The JAN (Joint Army/Navy) reliability levels are designated as JAN, JANTX, JANTXV and JANS. They may be abbreviated as J, JX, JV AND JS.

- JAN (J) - This designates qualified high-reliability parts listed in the Qualified Parts List (QPL). Parts in this group have undergone specific electrical, mechanical and environmental stress tests and pre-conditioning to a specified screening or reliability level. The testing is done on a randomly sampled size lot, as specified in the slash sheet, with a defective percentage of the lot allowable for lot acceptance. The maximum allowable percent of defective components (maximum number of defects per 100 units) that are considered satisfactory for the purpose of the sampling tests is called the Acceptance Quality Level (AQL).

- JANTX (JTX) - This designation adds extra testing (TX) of JAN parts with a higher screening level (generally done on a 100% testing level). All tests and pre-conditioning steps are either for a longer duration or greater stress, or both.

- JANTXV (JV) - This designation adds visual testing (V) to the JANTX parts with the addition of 100% visual inspection, using both microscope and X-ray techniques.

- JANS (JS) - This designation is for the highest level (S) of screening of the JANTXV parts with even more stringent tests and inspection methods specified.

# MILITARY STANDARD MIL-STD-750

This military standard provides the "how-to" test procedures for MIL-S-19500. It establishes uniform methods for testing all discrete semiconductor devices. This standard covers physical, electrical and environmental tests to determine the device's resistance to the detrimental effects of natural elements in conditions surrounding military operations. Suitable laboratory conditions are specified to simulate actual service conditions existing in the field for the rated life of the device.

MIL-STD-750 is categorized into test classes which list a group of different methods for the electrical, mechanical and environmental tests for different types of semiconductors.

# QUALIFICATION TO MIL-S-19500

Before a semiconductor manufacturer can ship any JAN devices, regardless of their reliability level, a qualification cycle must be performed. The qualification program is used to provide the assurance that the design, manufacturing process, assembly, inspection and testing of the semiconductor complies with the MIL-S-19500 specification and its applicable slash sheet. For example, MIL-S-19500/1N914 contains the detailed specifications for the JAN1N914 diode.

PRODUCTION LINE QUALIFICATION
The fabrication, assembly and testing lines used for JAN devices must be officially certified by a DCAS inspector for MIL-S-19500 use. Facility qualification specifically includes the following:

- Process control
- Facility cleanliness
- Equipment calibration
- Documentation

The certification process is as follows:
- The first step is a formal request for a certification audit.

- Once the DESC audit has been completed, a letter is sent to the manufacturer granting facility approval or outlining any required corrective actions.

- After the corrections have been made, the manufacturing facility would be reaudited and given a facility approval.

## DEVICE QUALIFICATION AND REQUALIFICATION

- Once product line certification has been granted, a specified lot size of the required semiconductor is manufactured for qualification testing. The qualification testing sequence includes screening and conformance to the specific sections of MIL-S-19500, referred to as Groups A, B and C. These tests, described later, are performed on a randomly selected sampling from the lot.

- After submission of the required test reports to DESC showing that all tests and inspections have been satisfactorily completed the manufacturer and that specific semiconductor will be listed in the Qualified Parts List (QPL) - 19500.

- To retain the QPL listing, each year, the manufacturer must submit a summary of the quality conformance testing that has been performed for a new lot of that device or for a device that is structurally identical.

- DESC, and not the final user, must be notified by the qualified manufacturer of any change in the product or its product assurance plans, if that change might affect performance, quality, appearance, reliability or interchangeability.

- DESC will then inform the manufacturer regarding the steps, if any, that must be taken to retain qualification. These steps may include a full qualification test sequence.

- After the manufacturer has conformed to the required steps, DESC will then review the newly submitted data and rule on allowing the changed product to be listed in the QPL.

The only parts that can be designed into equipment intended for military or aerospace applications must come from the Qualified Parts List, unless a condition exists where a desired device from a manufacturer is not listed in the QPL. Under this condition, permission to use the unlisted device must be obtained from the prime contractor who, in turn, must obtain approval from DESC.

## QUALITY CONFORMANCE TESTING

All levels of reliability require continual testing to be sure that the device conforms to the requirements of the military specification, MIL-S-19500, and to any additional details that may be covered in the slash sheet.

The requirements for processing each of the reliability levels will vary from level to level. The first level (JAN) does not require one hundred percent testing. Successful completion of a qualifying conformance test sequence that is run on a randomly selected sample of devices from the inspection lot qualifies the entire lot.

The JANTX, JANTXV and JANS reliability levels do require one hundred percent testing on a lot-by-lot basis. The devices conforming to the JANS level require wafer lot inspection and traceability, particle impact noise (PIN) detection, external visual inspection, radiography tests and serialization. The specific requirements for processing includes all the preconditioning steps, and mechanical and electrical testing that is designed to weed out weak devices with potential failure mechanisms.

# THE ELEMENTS OF MILITARY SPECIFICATION MIL-S-19500

## GROUP A TESTING
- These are the electrical tests listed in the manufacturer's data sheet. Tests are listed in the applicable slash sheet and are performed on a randomly selected sample taken from each inspection lot. When no characteristics are specified, no tests are needed to satisfy Group A requirements.

- Visual and mechanical inspection of the package and its leads are performed per MIL-STD-750.

- Depending on the screening level required, different sampling sizes are listed. Some of the tests for the JANS reliability level must be performed in a specific sequence within a particular sub-group test procedure.

These tests are considered non-destructive and parts may be shipped after being subjected to Group A tests.

## GROUP B TESTING
This group of tests is a die and package stress test sequence performed on a randomly selected sampling from each inspection lot. The tests are often referred to as the "shake, rattle and roll" tests and are used to verify proper wafer fabrication, chip assembly, lead bonding and package sealing. As with the Group A tests, each level of reliability has its own limits and methods

specified in MIL-STD-750, with tests classified under different sub-groups. They include the following:

● Lead solderability and resistance to solvents.

● Thermal shock (temperature cycling) - temperatures and cycle time are specified for each reliability level. Devices are initially placed in a high temperature oven and, when stabilized at that temperature, are then plunged into a cold environment. This test is used to verify lead bonding.

● Hermetic seal test - verifies the seal between the unlidded package and its cover. Depending on the required reliability level, either a gross or fine leak test is performed.

● Stabilization bake - this pre-conditioning is done at a specified high temperature and duration and serves to stabilize the electrical parameters of the device.

● Power burn-in - Devices are placed in an oven at a specified temperature and duration and operated at 80% of their maximum power capability. The values of the temperature and time are determined by the required reliability level. The burn-in procedure will expose any parts that have been improperly processed during wafer fabrication and serve to eliminate bad parts during this early period of device pre-conditioning. At the completion of power burn-in, changes are recorded from the results of the Group A tests of the specified electrical characteristics of the device. The slash sheet specifies the maximum changes allowed after burn-in.

● Decap and internal visual examination - A random sample group of the inspection lot is decapped and visually examined to verify structural design of the device. When specified, a scanning electron microscope (SEM) test may be performed. Lead bond and die shear tests may also be performed when specified on the slash sheets.

The documentation for Group B tests is very extensive and expensive. When required, they are specifically referred to in the purchase order for the required parts.

## GROUP C TESTING
These tests are used for qualification data to verify package integrity and include testing of the device under mechanical,

thermal and environmental stresses. All tests are run on a randomly selected sample with identical tests done for all four reliability levels. The following tests are performed by the specified methods of MIL-STD-750 except where noted.

- Electrical measurements as specified on the data sheet.

- Physical dimensions per the data sheet.

- Terminal strength and hermetic seal for gross and fine leaks.

- Mechanical shock, vibration, acceleration and corrosion in a salt atmosphere.

The last three tests may be done on electrical rejects from the same lot. Lot integrity may be confirmed by decapping of the defective parts after the tests are performed. Group C tests, however, are frequently waived and are so specified in the device slash sheet.

# QUALITY CONTROL PROGRAM

Through their establishment of standardized test methods, preconditioning processes and stringent requirements, military specification MIL-S-19500 and military standard MIL-STD-750 have provided an extremely effective technique for obtaining highly reliable discrete semiconductors. To standardize quality and inspection criteria, DESC has established three standard specifications:
- Quality Specification: MIL-Q-9858
- Inspections Specification: MIL-I-4520
- Calibration Requirements: MIL-C-45662

These specifications deal with the principles of maintaining a proper quality control system to assure compliance with the requirements of a military contract. The program used to implement these specifications is developed by the contractor. It must include proper documentation and is subject to review by the Federal Government.

An adjunct specification is the military standard MIL-STD-105. This standard defines the sampling levels used for Acceptance Quality Level (AQL) inspection (Normal or Tightened) or 100% inspection of incoming parts. The selected level is determined by the required level of reliability.

Much of the success of these procedures is dependent on the control exercised by both the component and the equipment manufacturer's Quality Control and/or Quality Assurance Departments.

## AT THE COMPONENT MANUFACTURER'S FACILITY
A satisfactory Quality Control program must assure adequate monitoring of the quality of all areas of the semiconductor manufacturing process. This includes the following:

- Design
- Development
- Fabrication
- Assembly
- Processing
- Visual inspection
- Electrical testing

- Equipment calibration
- Maintenance
- Sorting and branding
- Final packaging
- Storage
- Shipping
- Site installation

## AT THE EQUIPMENT MANUFACTURER'S FACILITY
Despite careful control at the semiconductor manufacturer's facility, some percentage of all devices shipped may be defective. Inconsistencies in production process control may occur during manufacture, testing, sorting and handling. This may result in the user receiving an inordinate percentage of defective parts. Electrical characteristics of the device may also change after production testing, even when done on a one hundred percent basis. This shift in characteristics is particularly critical in devices being used under conditions close to its specification limits.

To assure the user that the device intended for military or aerospace applications will operate properly, a carefully planned incoming inspection program is a necessary step prior to any assembly stage. Incoming inspection of all parts, especially semiconductors, will serve to minimize failure in sub-assemblies, major sub-systems, final products and, even more important, failures in the field. Preferably, incoming inspection should be performed on a 100% basis. At the very least, it should be done on a sampling basis.

# CAUSES OF SEMICONDUCTOR FAILURE

In any discussion of semiconductor failure, it is necessary to separate the causes of failure into two categories:
● Manufacture-related failure
● User-related failure.

## MANUFACTURE-RELATED
The establishment of proper process control and quality control techniques and adherence to MIL guidelines will eliminate, or minimize, the defects that are primarily due to a deficiency within the manufacturing procedure.

Semiconductor processing can be divided into four distinct stages of production. These stages are:

● Wafer fabrication   ● Assembly   ● Packaging   ● Testing

Failures can occur at any stage of the manufacturing process.

| FAILURE MECHANISM | CONTROL TECHNIQUES |
|---|---|
| **WAFER FABRICATION** | |
| Die metallization or oxide defects | Pre-cap visual, including SEM inspection (Scanning Electron Microscope) |
| Surface contamination | Stabilization bake |
| Poor photolithographic technique | Power burn-in |
| **ASSEMBLY AND PACKAGING** | |
| Poor die attach | Pre-seal visual |
| Improper lead bond | Thermal shock |
| Sealing defects | Hermeticity tests |
| **TESTING** | |
| Non-calibrated test equipment | Frequent, periodic inspection and calibration of equipment |
| Human or machine error: | |
| Improper marking | Tighter AQL sampling |
| Incorrect results | |
| Grading errors | Proper training and supervision |

USER-RELATED

1. When breakdown voltage ratings are exceeded because of a voltage transient or a sudden increase in supply voltage, the difficulty may be rooted in the improper selection of a device with a marginal breakdown voltage rating for its unique application.

2. Circuit failure due to polarity reversal of the system voltage supply

3. Power dissipation of the device is exceeded because of:
   - Inadequate power derating at elevated temperature

   - Insufficient heat transfer from the device because of:
     Poor thermal interface between case and heat sink caused by insufficient torque and/or lack of thermal compound at interface
     Inadequate or improper heat sink
     Fan or blower too small or non-existent
     Inadequate or no heat-venting holes in the enclosure

   - Heat build-up inside enclosure not accounted for in the circuit design

4. Components assembled into the circuit incorrectly

5. Excessive current because of inadequate circuit protection against a short in another component or breakdown of the insulation in the circuit

6. No protection against lightning or electrostatic discharge

7. Broken external terminal leads because of:
   - Improper handling of the package
   - Excessive bending of the leads
   - Inadequate strain relief on the lead-to-case seal when assembled into the circuit
   - Incorrect techniques for insertion of the package into the circuit

8. Contamination of the chip inside the package because of a cracked seal between case and leads. This can be caused by:
   - Improper handling of the package
   - Excessive strain on rigid terminal leads

9. Fracture of the chip due to mechanical distortion of package

# GENERAL COMMENTS ON HIGH-RELIABILITY SEMICONDUCTORS

Military and aerospace applications dictate the need for the most stringent of high-reliability requirements and quality assurance.

In the harsh environment of space and for the extreme demands of modern military electronic technology, the possibility of contamination and stress from external sources necessitate the controlled production of semiconductors and the equipment into which they are assembled. The components and the equipment must withstand the stresses and still operate within their required specifications for a specified length of time.

The term "high-reliability" does not necessarily mean an ideal or "perfect" component or a "perfect" piece of equipment. The achievement of perfection is either technically not feasible or too expensive.

What high-reliability does imply is the establishment of a pre-determined level of quality that will provide the confidence that the semiconductors, and eventually the circuits in which they are assembled, will operate according to the contracted specifications for a pre-determined duration.

CHAPTER
NINE

# TRENDS FOR DISCRETE SEMICONDUCTORS

POWER MOSFETS

POWER MOSFET PACKAGING

SURFACE MOUNTED DEVICE (SMD) TECHNOLOGY

# TRENDS FOR DISCRETE SEMICONDUCTORS

## POWER MOSFETS

Power bipolar transistors have long been considered the mainstay of the power segment of the discrete semiconductor industry. With new developments in manufacturing technology, power MOSFETs have been brought from obscurity in the 1970s to the point where they have become a major force as a discrete power component, particularly in the following applications:

- Regulated switching power supplies for computers, telecommunications and instrumentation equipment

- Automotive electronics
- A.C and D.C. motor controls

- Switching circuits
- Relays and solenoid drivers

The industry is projecting that power MOSFET technology will eclipse power bipolar technology by the end of this decade. The reasons for anticipating that power MOSFETS will dominate the discrete power transistor industry can be attributed to the following factors:

- Power MOSFETs have greatly improved in their power handling capability. Until recently, power bipolar transistors had much higher power capability at both low and high voltage levels because of their low ON resistance when operating as a switch. Improved MOSFET technology has produced devices with higher power capability and significantly lower values of $R_{ON}$ in devices with breakdown voltage ratings below 300 volts. This has resulted in considerably lower power losses and, therefore, higher switching efficiencies.

- New power MOSFETs have power capabilities at 25°C ranging from 1 watt to 300 watts. For switching applications, ON resistance values for power MOSFETs range between 0.005 ohms to 0.100 ohms at drain currents between 10 to 50 amperes. The values for ON resistance of comparable power bipolar power transistors are still lower than that of MOSFETs, however, the gap is closing very rapidly and it is anticipated that this difference will eventually be eliminated.

- Higher operating frequencies - Unlike small-signal types, power bipolar transistors are typically limited to about 100 Kilohertz in normal operation, whereas power MOSFETs can operate above 1 Megahertz. This is particularly useful in switching power supplies where high frequency capability produces higher switching speeds, resulting in higher power supply efficiency. The use of smaller and lighter components (transformer, filter choke and filter capacitors) serves to reduce overall costs. Typically, speed of response of a power MOSFET is 50 to 100 nanoseconds, compared with the relatively slower speed of response power bipolar transistors, generally specified in microseconds.

- Power MOSFETs use simpler drive circuitry resulting in easier design, fewer parts, less space, less weight and lower costs.

- By connecting several MOSFETS in parallel, current and power capability can be increased. In the case of power bipolar transistors, paralleling more than two devices can be complex and expensive. The trend in the development of power bipolar transistors (and power SCRs) is to design larger chips that can satisfy higher power requirements. As die size increases, however, a point is reached where the available mounting area or the heat transfer capability of low cost packages is exceeded, requiring more expensive packages.

  Regardless of the additional cost of the package, a single high power bipolar transistor approach may be less desirable than several devices connected in parallel. A heat sink can perform its work more efficiently if the heat generated in the transistors is absorbed through several parallel paths. This is easily accomplished with several power MOSFETs, rather than through one power bipolar transistor.

- Power MOSFETs feature improved breakdown characteristics. They have less susceptibility to failures that are due to excessive voltage transients than do power bipolar transistors with comparable breakdown voltage ratings and die sizes.

- The average selling price of power MOSFETs has declined significantly, although power bipolar transistors are generally lower in price than power MOSFETs. Overall system costs must be considered before a proper choice can be made. In some cases, use of power MOSFETs will be a less expensive option if the cost of simpler drive circuitry, and smaller transformer, chokes and filter capacitors are included.

● Unlike the power bipolar transistor, the MOSFET contains a "free" inherent PN junction as part of its structure. The equivalent circuit can be drawn as a MOSFET with a diode in parallel with its channel. See Figure 87.

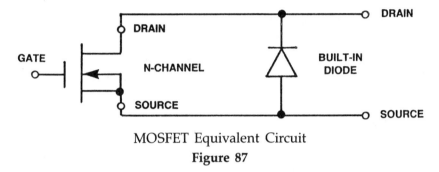

MOSFET Equivalent Circuit

**Figure 87**

This inherent internal diode can sometimes be used as a clamp to protect against the transient energy generated by switching an inductive load. See Figure 88.

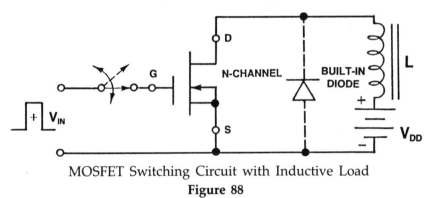

MOSFET Switching Circuit with Inductive Load

**Figure 88**

The counter EMF generated by switching an inductive load will reverse the drain-to-source voltage and cause the diode to conduct. This clamping action prevents undue voltage stress on the channel. Forward current and breakdown voltage ratings of the inherent diode are the same as the ratings of the MOSFET.

Although power bipolar transistors are well established and will not disappear from the scene, power MOSFETs are offering a significant threat to the dominance of power bipolar devices, particularly in the lower voltage, higher frequency (higher switching speed) applications in existing and new circuitry.

# POWER MOSFET PACKAGING

The move in power MOSFET technology toward higher frequencies, lower ON resistance and higher power capabilities is resulting in the development of new, low-cost plastic packages for use in commercial, industrial and consumer power circuits and for new, hermetically-sealed high-reliability packages for military and aerospace applications.

Historically, the hermetically-sealed TO-3 (now termed TO-224) and the plastic TO-220, have been the high-volume workhorses of the discrete power semiconductor industry. See Figure 89.

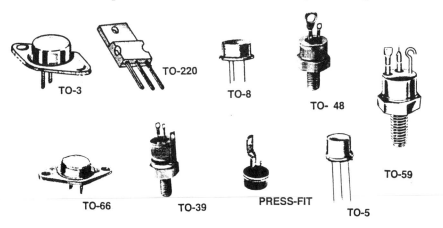

Existing Power Transistor Packages

**Figure 89**

The prefix "TO" in the package number stands for "transistor outline". This designation was assigned by JEDEC as part of the industry's standardization program.

These existing packages, however, do not provide the optimum means for using many of the features of the latest power MOSFET chips. New packages and packaging methods are being developed to provide higher power capacity, electrical isolation of the chip from the case, improved frequency and speed response and better heat transfer capability.

Until totally satisfactory packages for high power MOSFETs are standardized, most manufacturers are still offering their high power devices in the traditional package styles: TO-3 (TO-204), TO-220, TO-237 and TO-39 styles.

## COMMERCIAL, INDUSTRIAL AND CONSUMER TYPES

The TO-218, which resembles an oversized TO-220 package, has become the preferred choice for the newer power MOSFETs used in the high-frequency, high-speed switching power supplies and motor control applications. It was designed to meet the spacing and isolation VDE (Verband Deutsche Electrotechniker) standards for power transistors used among commercial European companies. It has wider physical lead spacing (.210" typically, compared with the .100 " for the TO-220) and longer distances along the critical package surfaces.

The TO-247, sometimes called the TO-3P, offers an integral plastic shoulder washer to simplify mounting the device tab to the surface of a heat sink. It still requires beryllium oxide, mica, or other isolation barriers to obtain good electrical isolation between package and heat sink.

A fully isolated version called the ISOWATT TO-18, has been introduced by several leading power MOSFET manufacturers. This package appears to offer the most economical solution to the problems of physical spacing, high frequency needs and total thermal management. At the same time, it provides electrical isolation without too severe a loss in thermal conductivity.

The 5-lead TO-220 provides a complementary pair of N-channel and P-channel power MOSFETs for compactness in heat sink requirements and reduced component count.

## MILITARY AND AEROSPACE TYPES

For military and aerospace applications, the TO-3 and TO-39 are still being used for JAN-qualified power MOSFETs, however, some special packages originally used for power bipolar transistors are now being used for the new power MOSFET chips. The isolated stud mount of the TO-61 contains a thermally-efficient (good heat transfer characteristics) beryllium oxide insulating wafer inserted between the chip and inner surface of the case. This package has become the preferred choice in airborne and high altitude applications. The thermal capability of the TO-61 package is presently superior to all other isolated types.

The very high current region is a major area in which hermetically-sealed packages used for military and space equipment still lags behind the capability of the high power chip. Because of the potential capability offered by power MOSFETs with very low $R_{ON}$, the present deficit in packaging capability becomes critical. Manufacturers of hermetically-sealed packages capable of handling chips with values of $R_{ON}$ less than 0.01 ohms are beginning to address this problem. Although some of the conventional packages are being used for the 100 to 300 ampere capability of these chips, an undesirable thermally-induced increase in $R_{ON}$ will result. The lowest possible thermal resistance will be the obvious benefit in using a package specifically designed to handle these very high currents.

With the high-reliability packaging required for high voltage power MOSFETs, pin-to-pin distances will become very critical as devices with voltage ratings in excess of 800 volts become more common. This could be particularly important in satellite and airborne applications where high voltages are normally used. In all likelihood, these packages will use advanced ceramics, or other appropriate materials, to provide a solution to pin-to-pin arcing problems.

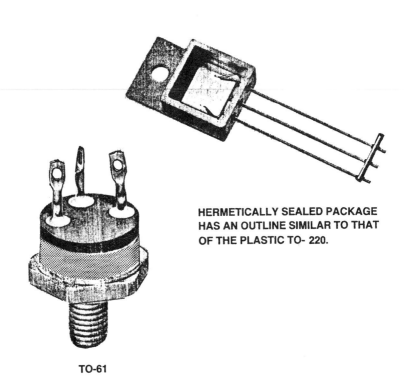

**HERMETICALLY SEALED PACKAGE HAS AN OUTLINE SIMILAR TO THAT OF THE PLASTIC TO- 220.**

**TO-61**

Military Packages

# SURFACE MOUNTED DEVICE (SMD) TECHNOLOGY

A new packaging technology in the form of surface mounted device (SMD) technology is responding to the demand for circuit miniaturization and lower printed board assembly costs. In Volume One of this series, this technology was examined with respect to its use with passive components and the placement and soldering techniques were described. This same packaging approach is being used for discrete semiconductors.

Many major semiconductor manufacturers are supplying many varieties of
diodes in pellet form on tape and reel.
They are being assembled with automatic placement equipment and soldered to both sides of a printed circuit
board.

The more complex multi-leaded discrete devices - SCRs, TRIACs, bipolar and field-effect transistors are provided in small packages applicable for surface mounting. The three-lead small outline packages, SOT-23 and SOT-89, are used for SMD mounting.

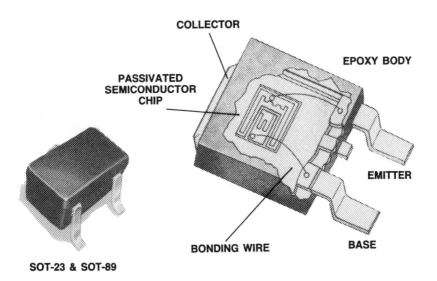

SOT-23 & SOT-89

Low-power SOT Packages For SMD Mounting

The physical size of the SOT-89 accommodates a chip about 350% larger than that used in the SOT-23 package that has a maximum power capability of 0.25 watts. The power dissipation of the SOT-89 package is a maximum of 1 watt, however, because of its extremely small size, the current rating of a device assembled in this package is severely restricted to relatively low values.

For higher power SMD devices, SOT packages are not used despite new chip technologies producing lower and lower values of $R_{ON}$. Only slightly better current ratings are possible because of their package-related restrictions.

The compromise options offered for power devices above 3 watts by the discrete semiconductor manufacturers are the following:
- A cropped and lead-formed version of the TO-220 package - with its leads bent either in a gull-winged or J-lead shape. Power capability is limited to the free-air package rating of 3 watts (or less). If the tab is soldered to an enlarged printed circuit foil, the power rating can be increased somewhat. This approach, however, only adds to the problem that SMD technology is meant to eliminate-the area used by large components.

- The lead-form version of the four-leaded TO-126 dual in-line package (DIP) offers a slightly smaller size package. With the option of forming the leads in either a gull-winged or J-lead style, use of this package takes less board space, provides easy visual inspection of solder joints and has the power capability of 6 watts and possibly higher.

These options are not the most desirable for the enhancement of SMD packaging techniques. At the present time, the conditions for sufficient heat transfer with the use of power devices and the amount of space taken by the large power packages are working against their use in SMD circuitry.

# ANSWERS TO REINFORCEMENT EXERCISES

# CHAPTER ONE - DEFINITIONS AND GENERAL INFORMATION

1. True

2. True

3. False - The element of time does not enter into this definition. A solid-state device is called a semiconductor because it can either act as a conductor when the proper voltage is applied or act as a nonconductor when a different voltage is applied, regardless of the voltage duration.

4. False - An integrated circuit is not a discrete device even though it is a semiconductor. All the other devices listed are classified as discrete semiconductors, since each is a single component in a finished package.

5. True - Even though the power dissipation capability of the device will be reduced as the temperature increases above 25°C and may go below 3 watts at high temperature, the specified rating at 25°C is used as the reference for classifying the device as a power or small-signal unit.

6. True

7. False - All these statements are true of silicon, not germanium. For most semiconductor devices that are made today, germanium no longer is used.

8. True

9. True

10. True

11. True

12. False - The opposite is true. Typically, a mechanical switch responds in hundreds of milliseconds. When used as a switch, an ordinary semiconductor will respond in hundreds of nanoseconds. A fast semiconductor will switch from ON to OFF, and vice versa, in less than 50 nanoseconds.

# CHAPTER TWO - DIODE PRODUCTION

1. True

2. True

3. True

4. False - Very few, if any, semiconductor manufacturers are involved in the processing of silicon dioxide to produce a silicon wafer. They generally purchase the raw silicon wafers from companies who only manufacture wafers. These wafers are then processed by the semiconductor manufacturers into finished semiconductor devices.

5. False - The technology of silicon rod and wafer production has progressed to the point where six inch diameter rods are quite common for use in the production of semiconductors. Some wafer manufacturers are even producing silicon rods that are eight inches in diameter.

6. True

7. False - The opposite is true. The use of boron as a dopant produces P-type material with the phosphorus dopant producing N-type material.

8. True

9. False - The stripe that is painted on the diode body is used to indicate the cathode terminal. In some cases, the diode symbol is branded on the body instead of a cathode stripe. The arrowhead is the anode and the bar is the cathode.

10. True

# CHAPTER THREE - DIODE CHARACTERISTICS AND SPECIFICATIONS

1. True

2. False - Just the opposite is true. When the diode is forward biased (positive voltage applied to the anode), it acts as a conductor. When reverse biased (negative voltage applied to the anode, or positive to the cathode), there is infinite resistance between the anode and cathode and the device is in a nonconducting state.

3. True

4. False - Decreasing temperature below 25°C does not increase its power capability. At temperatures below 25°C, the device will have the same power rating that is specified at 25°C.

5. False - If the product of the breakdown voltage times the reverse current through the diode is less than the rated power of the device, the device will not be injured in any manner. Under this condition, the reverse current is limited by the value of the resistor in series with the diode. Reducing the reverse voltage to a value less than breakdown will again put the diode into its nonconducting mode.

   If, however, the resistor is decreased in value so that the resulting increased reversed current multiplied by the value of the breakdown voltage produces a power greater than the rating of the device, the diode will be destroyed. Care must be taken to limit the reverse current with sufficient resistance if the peak inverse voltage is to be exceeded.

6. True - The ideal condition for a switch is when the forward resistance, $R_F$, is zero and the reverse resistance, $R_R$, is infinite. The front-to-back ratio would then be zero. Although a diode is not an ideal device, its front-to-back ratio is close enough to zero to be an effective switch.

7. True

8. True - When the current through the diode is at the 90% point of the final forward current, the diode is considered to be at its stabilized conducting state.

9. True

10. True

11. True - A zener diode may be put into its forward biased mode if desired, however, there is no useful application for the device in this mode. The purpose of the zener diode is to operate in its avalanche or breakdown region to maintain a predetermined constant voltage. Any load in parallel with the zener diode will also be maintained at the same constant voltage.

12. False - The dynamic zener impedance of a zener diode is defined as the change in reverse voltage as a function of a change in reverse current. If the voltage in the zener or avalanche region is constant, there is no change in reverse voltage as reverse current changes and the dynamic impedance of the zener is zero.

13. True

14. False - The "1N" prefix JEDEC designation indicates that the device is a semiconductor with two active terminals. In this case, this component is a diode with an anode terminal and a cathode terminal. The number after the prefix designation is assigned by JEDEC in sequential order. Other semiconductor components having three active terminals will be assigned a "2N" prefix.

15. True

# CHAPTER FOUR - DIODE APPLICATIONS

1. True

2. True

3. True

4. True

5. False - Because of the photo-voltaic effect, no external source of electrical power is necessary to produce light current when a light source impinges on the silicon.

6. False - As the reverse voltage across the varactor is increased, the value of the capacitor is decreased.

7. True

8. True

9. False - The reason for using a four-diode bridge instead of a dual-diode configuration for full-wave rectification is that there is no center-tap connection at the secondary of the power transformer. The bridge rectifier can be used with a center-tapped secondary winding, but no connection is made to the center-tap terminal. If only half of the secondary voltage is desired, the bridge is connected between the center-tap terminal and either end terminal on the secondary winding.

10. True - The combination of highest voltage, current and temperature all occurring simultaneously is referred to as the "worse case condition". This condition must be taken into account for proper circuit design.

11. True

12. True

13. True

# CHAPTER FIVE - THYRISTORS

1. True

2. True

3. False - The SCR will turn OFF when anode voltage is reduced to zero, regardless of the condition of the gate current.

4. True

5. False - To turn the SCR ON, gate current must flow at the same time that the positive voltage exists at the anode.

6. True

7. False - For certain applications, e.g. using the SCR as a latching relay, the supply voltage for the SCR is D.C.

8. True

9. False - The supply voltage for the gate circuit is generally obtained by connecting an RC series network across the A.C. power line. The junction point of the RC network is connected to the gate terminal to supply a forward bias voltage to the gate.

10. False - The output terminals of the TRIAC are designated as Main Terminal 1 and Main Terminal 2, not anode and cathode.

11. True

12. True

13. True

14. False - Increasing the torque beyond the specified amount may cause the package to distort and the thyristor chip to crack. Care must be taken in mounting the package not to exceed the manufacturer's torque recommendations.

# CHAPTER SIX - BIPOLAR TRANSISTORS

1. True

2. True

3. False - The PNP transistor requires negative voltages at its collector and base terminals with respect to the emitter and positive polarity voltages for the NPN transistor.

4. True

5. True

6. False - Power dissipation is specified on a data sheet under its absolute maximum ratings at a case temperature of 25°C. This value must be derated as temperature increases. This is true for all transistors regardless of rated power capability.

7. True

8. False - The current gain ($\beta$) of a bipolar transistor increases from zero as the collector current increase from zero. Depending on the transistor, it peaks at some value of $I_C$ and as collector current increases, beta decreases.

9. False - The cutoff frequency is defined as that frequency where the beta, at a specified value of collector current, has decreased to .707 of the constant beta value at a lower frequency. The selected transistor should be operated well below the frequency cutoff (before beta decreases).

10. True

11. True - When operated as a switch, the collector-to-emitter saturation resistance is essentially zero during the ON state, therefore, its power dissipation is essentially zero. When operated as an amplifier, power dissipation is higher since the collector-to-emitter resistance is a finite value.

12. False - The input resistance of a forward biased transistor is relatively low, loading down preceding circuitry.

# CHAPTER SEVEN - FIELD-EFFECT TRANSISTORS

1. False - The major feature of a field-effect transistor is its extremely high input resistance characteristic, causing no loading effect on preceding circuitry.

2. True

3. True

4. False - The built-in diode (PN junction) at the input section of the JFET must be at zero voltage or reversed biased to produce an extremely high input resistance.

5. True

6. False - The JFET is mainly used as a high input resistance linear amplifier, oscillator or voltage-controlled resistor.

7. True

8. True

9. True

10. True

11. True

12. True

13. False - The opposite is true. When zero voltage is at the input, the output is equal to $V_{DD}$. When a step voltage is applied to the input, the output is at zero volts.

14. False - In a CMOSFET switching circuit, either one of the series MOSFETs is always OFF with its channel resistance very high. This acts to minimize its load current and power dissipation. The drain-to-source voltage of the turned ON MOSFET is essentially zero, therefore, it is dissipating essentially zero power.

15. True

# GLOSSARIES

COMMON ACRONYMS AND
MISCELLANEOUS DESIGNATIONS

POPULAR ELECTRONIC TERMS
DISCRETE SEMICONDUCTORS

ELECTRONIC SYMBOLS

# LIST OF COMMON ACRONYMS
# AND MISCELLANEOUS DESIGNATIONS

AQL - Acceptance Quality Level

CMOS - Complementary Metal-Oxide Semiconductor

DCAS - Defense Contract Administration Service

DESC - Defense Electronic Supply Center

DIAC - Diode A.C. switch

DO - Diode Outline

EPI - Epitaxial Layer

ESD - Electrostatic Discharge

FET - Field-Effect Transistor

GSI - Government Source Inspection (or Inspector)

JAN (J) - Joint Army/Navy

JANS (JS) - Joint Army/Navy - most Stringent reliability level

JANTX (JTX) - Joint Army/Navy with extra Testing requirements

JANTXV (JTXV) - JANTX with Visual inspection requirements

JEDEC - Joint Electron Device Engineering Council

JFET - Junction Field-Effect Transistor

MIL-C-45662 - Calibration Requirements

MIL-Q-9858 - Quality Specification

MIL-I-4520 - Inspection Specification

MIL-S-19500 - Military Specification-Discrete Semiconductors

MIL-STD-750 - Military Standard Test Methods for MIL-S-19500

MOSFET - Metal-Oxide Semiconductor Field-Effect Transistor

N-CHANNEL - Negative-type material Channel

NMOS - N-channel Metal-Oxide Semiconductor

NPN - Negative/Positive/Negative structure

P-CHANNEL - Positive-type material Channel

PIN - Particle Impact Noise

PMOS - P-channel Metal-Oxide Semiconductor

PNP - Positive/Negative/Positive structure

QPL - Qualified Parts List

SCR - Silicon Controlled Rectifier

SEM - Scanning Electron Microscope

SMD - Surface Mounted Device

SOT - Small Outline Transistor

TO - Transistor Outline

TRIAC - Triode A.C. switch

# POPULAR ELECTRONIC TERMS
## DISCRETE SEMICONDUCTORS

**ANALOG AMPLIFIER** - A circuit whose output waveshape is an amplified version or analog of its input waveshape. Also called a LINEAR AMPLIFIER.

**ANALOG SWITCH** - A digitally controlled switch that will provide a conductive path for a small linear or analog voltage during its ON state.

**ANODE** - One of the two terminals of a diode (positive type material) or the output terminal (also positive type material) of a silicon controlled rectifier (SCR).

**BASE** - The input terminal of a bipolar transistor.

**BETA** (β) - The Greek letter which is used to designate the current gain of a bipolar transistor. It is the ratio of the transistor's output current to its input current.

**BIAS VOLTAGE** - The D.C. voltage applied across the terminals of a PN junction (diode structure), whether the device is a diode, bipolar transistor or JFET. A PN junction is forward biased when a positive voltage is applied to the P-region with respect to the N-region, and reversed biased when the voltage polarity across the PN junction is reversed.

**BIPOLAR TRANSISTOR** - A three-terminal semiconductor device with a three-layer structure of alternate negative and positive type (NPN or PNP) materials. It is capable of providing gain or amplification in a circuit.

**BRIDGE RECTIFIER** - A configuration of four semiconductor diodes that act to change A.C. to full-wave pulsating D.C.

**CATHODE** - One of the two terminals of a diode (negative type material) or the terminal (also negative type material) that is common to both input and output sections of an SCR.

**CHIPS** - Unpackaged diodes, bipolar transistors, SCRs, TRIACs and field-effect transistors (FETs). Also called dice.

**CMOS (COMPLEMENTARY METAL-OXIDE SEMICONDUCTOR FET)** - A three-terminal MOSFET that combines an N-channel and a P-channel MOSFET in a single switching circuit. This circuit features very low power dissipation and the effective elimination of an external load resistor. The device responds to a digital pulse at its input by turning one section of the device ON and the other OFF, causing the turned OFF section to act as its high-resistance load. When the input pulse reverts to zero, the condition of the two sections of the device are reversed.

**COLLECTOR** - The output terminal of a bipolar transistor.

**COMPLEMENTARY TRANSISTORS** - A dual arrangement of NPN and PNP bipolar transistors or N-channel and P-channel FETs in which the polarity of one device is the reverse of the other. They are generally used as a matched pair with identical electrical characteristics.

**DIAC** - A two-terminal bidirectional semiconductor diode A.C. switch used for triggering a TRIAC.

**DIFFUSION** - One of a series of steps in the fabrication of a semiconductor. This step introduces a small amount of a chemical element, called an impurity or dopant, into the substrate. It will generally produce either an N-type or a P-type region to create a desired electrical characteristic.

**DIGITAL SWITCH** - A switching circuit that turns ON and OFF in a step-function manner in response to a step-function or digital pulse.

**DIODE** - A two-terminal semiconductor device that will allow current to flow through it in only one direction. With the proper voltage polarity across the device, it will act as a conductor. When the voltage polarity is reversed, the device will act as a nonconductor, allowing no current to flow.

**DOPANT** - See IMPURITY.

**DRAIN** - The output terminal of a field-effect transistor.

**EMITTER** - The designation for the terminal in a bipolar transistor that is generally used as the terminal common to both the input and output sections of the device.

**EPITAXIAL GROWTH (EPI)** - An optional step in the semiconductor manufacturing process in which the raw wafer is prepared for further fabrication steps. With silicon semiconductor devices, silicon is precipitated in gaseous form to grow on the surface of a silicon wafer.

**FET (FIELD-EFFECT TRANSISTOR)** - either junction (JFET) or Metal Oxide Semiconductor (MOS). It is a three-terminal semiconductor that is used either in an amplifier or switching circuit. The major characteristic of a FET is that it has an extremely high input resistance.

**FORWARD RECOVERY TIME** - The length of time required for a diode in its reverse biased (OFF) state to recover to a stabilized ON state after a digital forward bias voltage is applied. If this time is 50 nanoseconds or less, the diode is applicable for use in computer and/or high-speed logic circuits.

**FULL-WAVE RECTIFIER** - A configuration of either two or four diodes that acts to change A.C. to full-wave D.C. The two-diode configuration is used in conjunction with a center-tapped secondary of a transformer. The four diode configuration is used when no center-tap exists at the transformer secondary and is called a BRIDGE RECTIFIER.

**GATE** - The input or control terminal of an SCR, TRIAC or FET.

**HALF-WAVE RECTIFIER** - A single diode that acts to change A.C. to half-wave pulsating D.C.

**IMPURITY** - An element added to the base material (either germanium, silicon or gallium) in the semiconductor fabrication process to create a P-type or N-type section as desired. For germanium, the impurities are arsenic and bizmuth; for silicon, boron and phosphorus; for gallium, arsenic and phosphorus.

**ISOWATT** - Isolated version of the TO-18 package.

**JFET (JUNCTION FIELD-EFFECT TRANSISTOR)** - A three-terminal semiconductor device constructed with an effective diode at its input and a channel in the output section. The input section is operated under a reverse or zero bias condition to provide an extremely high input resistance. It is generally used as part of an amplifier circuit and sometimes as a digital switch.

**LINEAR AMPLIFIER -**
See ANALOG AMPLIFIER.

**MAIN TERMINAL 1 AND MAIN TERMINAL 2** - The output terminals of a TRIAC, alternately acting as either an anode or a cathode.

**MOSFET (METAL-OXIDE SEMI-CONDUCTOR FIELD-EFFECT TRANSISTOR)** - A three-terminal semiconductor device with a built-in capacitor at its input and a channel in its output structure. It has an extremely high input resistance characteristic and is either an enhancement type or enhancement-depletion type device. The enhancement type MOSFET is used as a normally-OFF digital or analog switch and the enhancement-depletion type is used as an extremely high input resistance linear amplifier.

**PINCH-OFF VOLTAGE** - The value of reverse bias voltage applied to the input of a J-FET to cut off its output current.

**PN JUNCTION** - This is the simplest semiconductor structure and as a discrete device is called a diode. It consists of a positive or P-region (excess positive ions) in junction with a negative or N-region (excess negative electrons).

**RECTIFIER** - A semiconductor diode, or group of diodes, that acts to change A.C. to pulsating D.C.

**SCR (SILICON CONTROLLED RECTIFIER)** - A silicon device with four layers (PNPN) having an input or control terminal, an output terminal and a third terminal that is common to both input and output. It is generally used to provide half-wave rectification of an A.C. supply voltage with a controlled conducting period for power control circuitry.

**TRIAC (TRIODE AC SWITCH)** - A three-terminal silicon device that functions as two SCRs configured in an inverse, parallel arrangement, providing a means of controlling an A.C. voltage. The TRIAC is used in A.C. lighting and heating control circuitry.

**VARACTOR DIODE** - A two-terminal semiconductor diode that acts as a variable capacitor as its reverse bias voltage varies.

**ZENER DIODE** - A two-terminal silicon semiconductor diode having the unique characteristic of providing a predictable value of voltage breakdown when in its reverse biased condition. In going into its breakdown mode, it exhibits a sharp break from its nonconducting state into its conducting or avalanche state, maintaining a constant value of reverse voltage across it. The zener diode is used as a voltage regulator, voltage reference and circuit protection device.

# GLOSSARY OF ELECTRONIC SYMBOLS

# GLOSSARY OF ELECTRONIC SYMBOLS

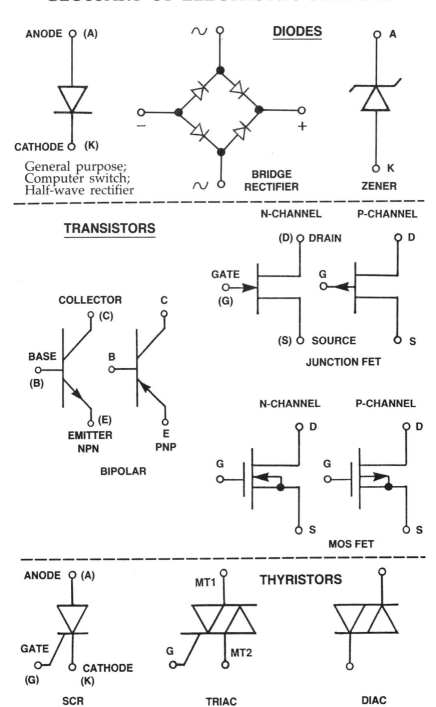

**DIODES**

ANODE (A)

CATHODE (K)

General purpose;
Computer switch;
Half-wave rectifier

BRIDGE
RECTIFIER

ZENER

A

K

**TRANSISTORS**

N-CHANNEL    P-CHANNEL

(D) DRAIN    D

GATE
(G)

G

(S) SOURCE    S

**JUNCTION FET**

COLLECTOR    C
(C)

BASE    B
(B)

(E)    E
EMITTER    PNP
NPN

**BIPOLAR**

N-CHANNEL    P-CHANNEL

D    D

G    G

S    S

**MOS FET**

**THYRISTORS**

ANODE (A)

GATE

(G)    CATHODE
(K)

**SCR**

MT1

G    MT2

**TRIAC**

**DIAC**

# PART TWO
# OPTOELECTRONICS

# INTRODUCTION TO OPTOELECTRONICS

During the early 1960s, the principles of the established sciences of optics and electronics and their resources were combined to create a new technology called *Optoelectronics*. This new technology dealt with the complementary phenomena of converting light into electrical current and/or converting electrical current into light. The scope of optoelectronics includes the devices, circuits, systems and materials used in the transmission and reception of light; it involves those circuits that control the electrical energy converted from light.

The term *optics* refers to the branch of science dealing with light and vision and covers two sources of light: *incandescence* and *electroluminescence*.

## INCANDESCENT LIGHT
An incandescent lamp consists of a tungsten filament mounted in a glass bulb from which air has been removed. The evacuated glass bulb is then filled with an inert gas, such as nitrogen, with its filament connected to external terminals. When either A.C. or D.C. voltage is applied across the terminals, a current will flow in the circuit creating intense heat in the filament, generating a bright glow called *incandescent light*. The emitted white light ranges over the entire light spectrum — from the nonvisible infrared, through the visible range, to nonvisible ultraviolet. Thomas Edison invented the incandescent lamp in 1879 by using a filament of carbonized bamboo mounted in a bulb from which air was evacuated to prevent oxidation. The development of more efficient techniques and materials has produced the improved tungsten filament lamp used today.

Although they are generally satisfactory in home and office lighting systems and a variety of indicator applications, incandescent lamps have their drawbacks. They can be broken by shock and vibration, use a great deal of power and their filaments burn out in about a year or so. This short life span makes incandescent lamps impractical for use as indicators and displays in electronic equipment. They are, however, suitable for use in large outdoor displays. Incandescent lamps are gradually being replaced by new technology display components that are more attractive, lower-powered and longer-lasting.

## ELECTROLUMINESCENT TECHNOLOGY

### ELECTROLUMINESCENT LIGHT

Electroluminescence is the light emitted from a solid or gas material over an extremely narrow section of the light spectrum. Although the bandwidth of the emitted light extends over several wavelengths, because of the sharpness (peak) of its waveshape, it appears as a single color light source. Light is emitted by energizing a solid or gas material with the application of the appropriate voltage or current.

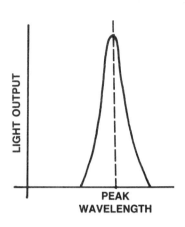

Components that emit electroluminescent light include: light emitting diodes, gas discharge and vacuum fluorescent tubes.

## LIGHT EMITTING DIODES (LEDs)

### VISIBLE LEDs

LEDs are available in four visible colors - red, orange, yellow and green for use in the following applications:
- Light indicators - single units and arrays.
- Segmented and dot matrix alpha-numerics digital displays
- Bar graphs and lighted bars

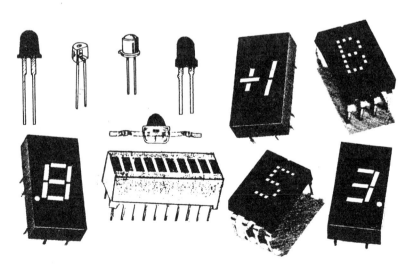

Visible Light Emitting Diodes (LEDs)

## INFRARED LEDs

LEDs emitting nonvisible infrared provide the light source in the transmitting section of fiber optic communications systems. They are also used in conjunction with other semiconductors (diodes, transistors and SCRs) to produce a wide range of opto-couplers (opto-isolators). Opto-couplers are used to isolate voltage levels between two circuits when transferring information between them and are sometimes used to minimize electrical noise when it exists in the circuit reference section.

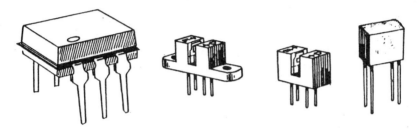

Opto-couplers

The *solid-state relay* (SSR), the semiconductor version of an electromechanical relay is an extension of opto-isolator technology. Using additional circuitry and an appropriate output, the SSR can interface low voltage digital logic circuits with high current, high voltage A.C. or provide relay action while eliminating the disadvantages of electromechanical switching.

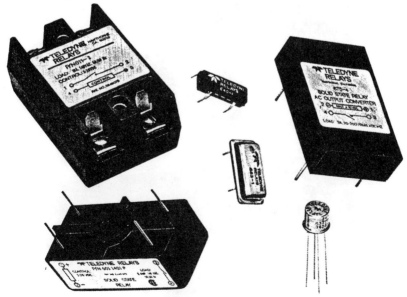

Solid-state Relays

# GAS DISCHARGE (COLD CATHODE OR PLASMA) DISPLAYS

Gas discharge (also called *cold cathode* or *plasma*) displays use an inert gas, such as neon, as the light emitting material. When a low voltage (or no voltage) is applied across the gas, the circuit has infinite resistance; there is no current flow and no light emission.

When a sufficiently high voltage is applied across the terminals of the selected elements of the display, the gas will ionize (breakdown), decreasing its resistance to allow current to flow, resulting in a bright orange display. Other colors such as green, red or amber are attained by changing the gas used in the display.

Gas discharge displays are used for:
- Segmented and dot-matrix digital displays
- Linear or circular bar graphs of all forms
- Light bars and flat-panel displays

Gas Discharge Displays

## VACUUM FLOURESCENT DISPLAYS

*Vacuum fluorescent* displays use phosphor-coated screens that are energized when a stream of electrons coming from a heated filament bombard the screen. Two types of displays are available — the *cathode ray tube* (CRT) and the *fluorescent indicator panel* display.

CATHODE RAY TUBE (CRT)
The familiar CRT is used as the picture tube in a TV set or as the video monitor in a computer or radar console. The color of the display, might be bluish-green, amber or white, depending on the type of phosphor used on the screen.

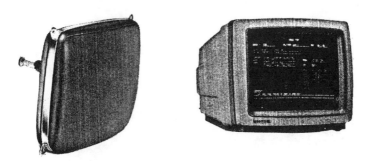

Cathode Ray Tubes

FLUORESCENT INDICATOR PANEL DISPLAYS
Fluorescent indicator panel displays are used for:
● Segmented and dot-matrix digital displays
● Flat-panel picture, charts and bar graph displays
● Special graphics displays

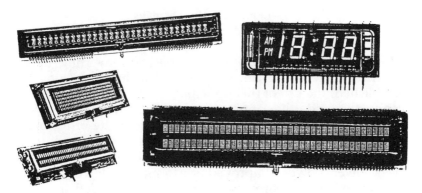

Fluorescent Indicator Panels

## LIQUID CRYSTAL TECHNOLOGY

The *liquid crystal display* (LCD) is neither incandescent nor electroluminescent in nature. Instead of emitting light, it scatters ambient visible light when a voltage is applied to a desired pattern etched on the display. The readout appears as a black or grayish (silver-toned) character, however, with the use of special filters other colors can be displayed. If no ambient light or backlighting is present, no display appears.

Of all display techniques, LCD technology uses the least power and is most suitable for battery-powered, portable operation. Liquid crystal displays are used in digital watches, hand-held calculators, portable computers, digital meters and telephones.

Liquid Crystal Displays

Although the optoelectronic principle of converting light to electrical current and electrical current to light has long been established, the use of this principle, in combination with semiconductors and other appropriate materials, made possible the manufacture of new and unique products. State-of-the-art optoelectronic components have been developed to work in systems having faster switching speeds, higher operating frequencies, lower power, improved performance, lower cost and greater reliability compared with systems intended for similar applications using the older, more traditional components.

 CHAPTER
TEN

# LIGHT EMITTING DIODES (LED)

# LIGHT EMITTING DIODES (LED)

Light emitting diodes (LEDs) are semiconductor diodes (PN junctions) that generate essentially a monochromatic light (single color) when a forward bias voltage is applied across their terminals. A reverse bias voltage, or zero voltage at the terminals, will turn the device OFF.

## LED CONSTRUCTION

The processes used to manufacture light emitting diodes are an outgrowth of the techniques used for the manufacture of silicon semiconductor devices. The combination of gallium, arsenic and phosphorus is used to form rods, or ingots, grown by the Czochralski method. See CHAPTER TWO — DIODE PRODUCTION.

- The rods are either gallium phosphide (GaP), gallium arsenide phosphide (GaAsP) or gallium arsenide (GaAs) from which wafers are sliced to be used as the substrate material.

- The wafers are mechanically and chemically polished to produce a crystalline structure which is virtually free of unwanted impurities and other imperfections.

- Nitrogen and zinc are used as the dopants to create the N and P regions. Crystal growing, epitaxial deposition, precise photolithography and carefully controlled impurity diffusions all contribute to the final LED chip product.

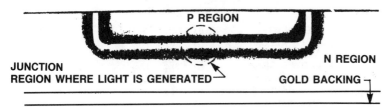

Cross section of a LED chip
**Figure 90**

- The completed diode chip is coated with a transparent silicon dioxide layer to improve light emitting efficiency by a factor of 2 or more. The chip is mounted on an appropriate package frame and encapsulated in a plastic or metal enclosure.

- The epoxy is tinted with a dye that absorbs ambient light and makes the LED appear darker when it is OFF. Since the dye does not absorb the radiated light from the chip, the viewing contrast (ON/OFF ratio) is improved.

- Gold backing on the bottom of the chip is used to reflect light that might be absorbed by the mounting surface, adding to the forward-emitted radiation of the junction area.

# LED PACKAGES

### T-1 AND T-1 3/4 PACKAGES
When light emitting diodes were first developed, the standard miniature-sized incandescent lamp packages, the T-1 and T-1 3/4 sizes were used. The number after the "T" designates the lamp diameter in eighths of an inch. See Figure 91.

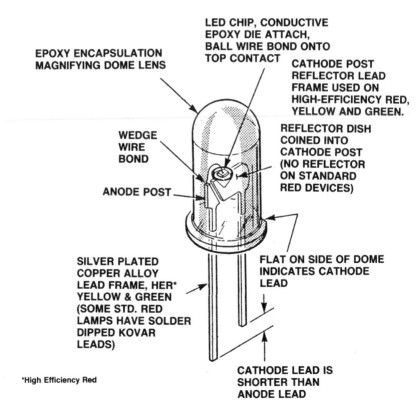

Construction Features of a T-1 3/4 Epoxy Encapsulated LED
**Figure 91**

The construction of the smaller T-1 package is essentially the same as the T-1 3/4. The reflector dish used in the T-1 3/4 is too large to be encapsulated in the T-1, therefore, a more compact lead frame, without a reflector, is used in the T-1.

With an increased expansion of LED applications, the standard T-1 and T-1 3/4 packages could not be easily adapted to meet new size and shape requirements. As a result, new LED packages have been developed to satisfy these needs.

## SUBMINIATURE PACKAGE

The smaller subminiature package is lower in cost than the T-1 allowing for high packaging density in an array assembly. Unlike the T-1 which has no reflector, a small reflector cup is formed by bending the sides of the die attach and bonding pads, producing light emission similar to the T-1 3/4. See Figure 92.

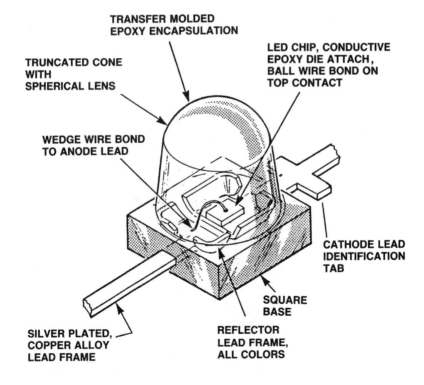

Construction Features of a Subminiature LED Lamp
**Figure 92**

The leads of the subminiature package may be left in the radial position for soldering to pads on the surface of a printed circuit board for SMD technology packaging, or the leads may be bent 90° into the axial position for insertion into the plated-thru holes on a conventional PC board.

The dome is a truncated cone with a spherical lens producing magnification for high light intensity. The package base is square-shaped, providing firm mechanical support for the lead frame and a means of positive package alignment in high density array applications.

RECTANGULAR PACKAGE
The rectangular package is most effective for backlighting a transparent legend or character. Bar graphs are also easily created with rectangular LED lamps that can be end-stacked or side-stacked to form a continuous bar of light. See Figure 93.

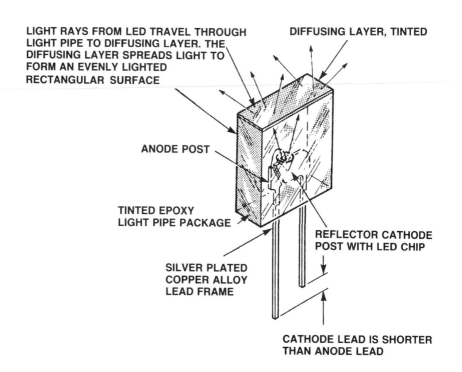

LIGHT RAYS FROM LED TRAVEL THROUGH LIGHT PIPE TO DIFFUSING LAYER. THE DIFFUSING LAYER SPREADS LIGHT TO FORM AN EVENLY LIGHTED RECTANGULAR SURFACE

DIFFUSING LAYER, TINTED

ANODE POST

TINTED EPOXY LIGHT PIPE PACKAGE

REFLECTOR CATHODE POST WITH LED CHIP

SILVER PLATED COPPER ALLOY LEAD FRAME

CATHODE LEAD IS SHORTER THAN ANODE LEAD

Construction Features of a Plastic Rectangular LED Lamp
**Figure 93**

The rectangular LED lamp is constructed much like the T-1 3/4 package. It uses the same reflector lead frame in a different encapsulating form. The encapsulation is a rectangular *light pipe* with a diffusing layer at the top to provide an evenly lighted rectangular source of light.

HERMETICALLY-SEALED TO-18 PACKAGE
A light emitting diode in a TO-18 package is used for military and aerospace equipment or in industrial applications where the LED may be exposed to hostile environments.

The package has an optical glass window hermetically sealed at the top of its cap with the addition of an epoxy dome to increase viewing angle and to provide good ON/OFF contrast. See Figure 94.

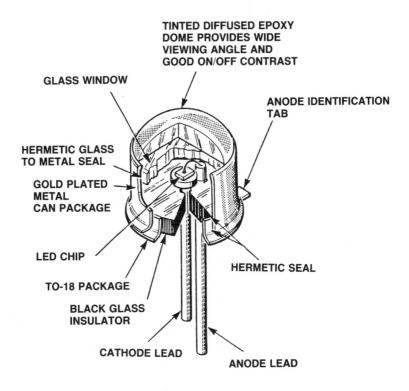

Construction Features of a Hermetically-Sealed TO-18 LED Lamp
**Figure 94**

# PRINCIPLE OF OPERATION

When a positive voltage is applied to the anode terminal of a light emitting diode with respect to its cathode, the device is forward biased, creating essentially zero resistance across its PN junction, allowing current to flow in the LED.

As the current enters the region near the PN junction, a packet of electromagnetic energy (*photons*) is released. The photon, having a specific peak wavelength, produces an essentially *monochromatic* (single color) light, with its intensity depending on the current flowing in the LED; the greater the current, the greater the light intensity. Current is determined by the supply voltage, E, and resistor, $R_S$, in series with the LED. Figure 95. Some LED packages include a resistor chip, eliminating the need for an external resistor.

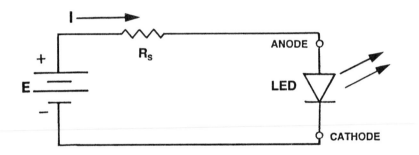

LED Circuit
**Figure 95**

The symbol for the LED is the same as that of an ordinary diode with the addition of two small arrows coming from the LED to indicate that the device can emit light.

The specific color of the light emitted depends on the substrate material that is used and on the concentration of dopants diffused into the substrate to produce the P-type and N-type regions. The emitted light is either one of the visible colors — red, orange, yellow, green — or nonvisible infrared.

The substrates used to produce these colors are:
- Gallium arsenide phosphide (GaAsP) - red, orange and yellow
- Gallium phosphide (GaP) - green
- Gallium arsenide (GaAs) - infrared

# LED SPECIFICATIONS

In addition to the same information contained in an ordinary diode data sheet, optical characteristics are included.

## PACKAGE OUTLINE
The drawing of the package outline appears on the data sheet and includes information on package dimensions, mounting details, lead designation, lead materials, color and type of lens (diffused or clear).

## ABSOLUTE MAXIMUM RATINGS

PEAK INVERSE (REVERSE) VOLTAGE - PIV or PRV (Specified at 25°C)
This rating specifies the maximum reverse voltage (negative voltage to the anode) that can be safely used and still have the LED maintain its nonconductive state. Typical values range between 3 to 25 volts. Generally, the LED is switched to its OFF state by removing the voltage to the diode.

CONTINUOUS D.C. FORWARD CURRENT - $I_{Fcont}$
Maximum allowable current for continuous D.C. operation. This value typically ranges from 50 to 100 milliamperes.

PEAK FORWARD CURRENT - $I_{Fpeak}$
The maximum allowable peak forward current for non-continuous operation. The rating is based on short duration pulse widths (normally 1 microsecond) at 300 pulses per second. Different pulse widths and repetition rates can be used. Typical values range from 1 to 5 amperes.

D.C. POWER DISSIPATION - $P_D$ (At 25°C case temperature)
The maximum power capability depends on the size of the LED chip, the type and size of the package and its mounting features. Typically, values range from 75 to 250 milliwatts.

## OPERATING AND STORAGE TEMPERATURE
For plastic devices, the range is from -55°C to +100°C. Hermetically sealed devices range between -55°C to +125°C.

Once a light emitting diode has been committed to a specific combination of materials, the color of its emitted light will be substantially the same, despite the minor effect of changing temperature. Increased temperature will cause a slight increase in wavelength (slight frequency decrease).

## ISOLATION VOLTAGE

The maximum voltage that can be applied between a lead of the LED and the package case. This value ranges from 300 to 500 volts.

## LEAD SOLDERING TEMPERATURE

Typical value is 230°C for 7 seconds with the soldering spot at least 1/4 inch from the seal between lead and package.

# PACKAGE MOUNTING

Because the estimated life of a light emitting diode is about 1 million hours (114 years), there should be no reason to replace the device if it is operated within its rated limits.

Unlike an incandescent lamp whose filament will generally burn out within a few years (requiring a lamp holder for ease in replacement), LED terminals can be permanently soldered into a circuit. LEDs can also be permanently mounted in many ways on a front panel or a printed circuit board as shown in Figure 96.

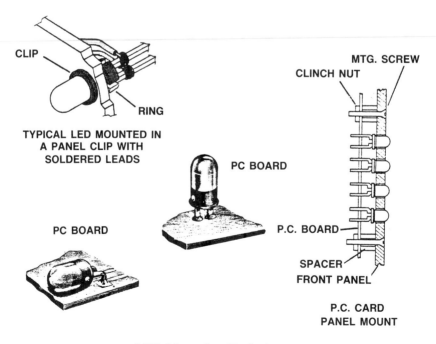

LED Mounting Techniques
**Figure 96**

## ELECTRICAL CHARACTERISTICS
At an ambient temperature of 25°C, unless otherwise noted

### FORWARD VOLTAGE ($V_F$)
This voltage value is the sum of the threshold voltage, $V_{TH}$, and the voltage across the resistance, $R_F$, of the LED. See CHAPTER THREE — DIODE CHARACTERISTICS AND SPECIFICATIONS for a discussion of $V_{TH}$ and $R_F$. The color (peak wavelength) of a light emitting diode is a function of its chip material, which also determines the forward voltage, $V_F$, of the LED. Typical forward voltages for the different colored LEDs are shown in Figure 97.

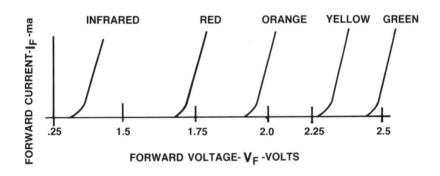

Typical Forward Voltages
**Figure 97**

### REVERSE LEAKAGE CURRENT ($I_R$)
This is the maximum reverse current measured at the peak reverse voltage value. Typical values are from 2 nanoamperes to 100 microamperes, depending on the specific device.

### CAPACITANCE
The value of PN junction capacitance is measured at zero voltage. Typical values range from 25 to 700 picofarads depending on the geometry of the chip and package size.

### SPEED OF RESPONSE
This is the time required to turn the device ON and OFF. See CHAPTER THREE — DIODE CHARACTERISTICS AND SPECIFICATIONS for a discussion of this characteristic. Typical speed of response values for LEDs range from 1.0 nanoseconds to 25 nanoseconds, depending on chip geometry and package size.

## OPTICAL CHARACTERISTICS
### (at 25°C ambient temperature)

PEAK WAVELENGTH (COLOR)
This characteristic is specified as a typical and maximum value at the peak of the radiated spectral response curve in either nanometers or Angstrom units. (One Angstrom unit is equal to 0.1 nanometers.) Although the bandwidth of the emitted light extends over several wavelengths, because of the sharpness of the waveshape, it appears as a single-color light source. Typical spectral response curves of the four visible colors (red, orange, yellow and green) and infrared are shown in Figure 98, with percent relative light output plotted against wavelength.

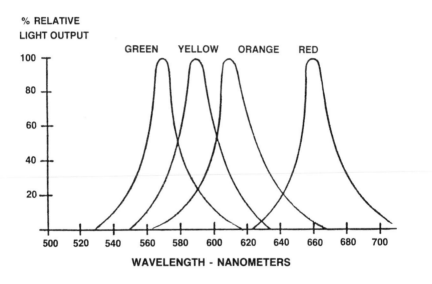

Spectral Response Curves of LED Light
**Figure 98**

LUMINOUS INTENSITY (I)
The intensity of the emitted light is specified at some value of forward current ($I_F$), in *candela* (cd), *millicandela* (mcd), or *microcandela* ($\mu$cd), however, *lumen* is sometimes used as the measurement unit. One candela is equal to 12.56 lumens. Luminous intensity is a measurement of the number of lines of light (luminous flux) that radiate from a light source in the form of a cone. Figure 99 shows luminous intensity as a function of forward current, $I_F$. Note that the change in luminous intensity is very nearly linear with a change in $I_F$.

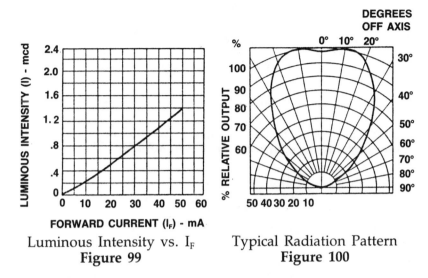

Luminous Intensity vs. $I_F$          Typical Radiation Pattern
**Figure 99**                              **Figure 100**

The radiation pattern (Figure 100) indicates the luminous intensity of the LED at different angles. Both the luminous intensity and the radiation pattern are necessary to properly specify the optical characteristics of a device.

# LED CHARACTERISTICS

- A light emitting diode is most applicable for battery-powered equipment, requiring very little power for normal operation — approximately 2 to 6 milliwatts. This is about one thousandth the power required for a typical incandescent lamp used as an indicator.

- The LED has low voltage and low current requirements. It needs about 2 to 3 volts at 1 to 2 milliamperes for normal light intensity when used as an indicator lamp. This feature makes the device particularly applicable for modern systems that require low voltage and low current power supplies.

- Except for the incandescent lamp, the operating temperature range for the LED is the widest of all display devices: from -55°C to +100°C for the plastic types and from -55°C to +125°C for hermetically sealed types.

- The change in luminous intensity of a light emitting diode is very nearly linear with a change in current. This characteristic is useful in systems using linear or analog control.

- Unlike other indicators and digital displays, the estimated life of LEDs is 1 million hours (114 years), the longest of all opto-electronic indicator or display products. Because of its longevity, the device requires no replacement if operated within its ratings and no additional lamp holder is needed.

- It is shock and vibration proof.

- Being a solid-state device, it has all the reliability that is inherent in semiconductor technology.

Historically, the major disadvantage of a light emitting diode was that its light could not be seen when exposed to direct sunlight. This condition, originally true, has been corrected with the latest LED product technology. High-efficiency, high-intensity red, yellow and green LEDs made from gallium phosphide material provide clear readability in direct sunlight. Yellow and green devices are used in avionic and automotive applications while the red LED is used as a warning color.

The contrast of the high-intensity LED in direct sunlight can be enhanced with the use of an appropriate mechanical light shielding technique to achieve optimized readability.

# LED APPLICATIONS

Visible light emitting diodes are used as:

- Single light indicators in simple visual indicator circuits

- Arrays with multiple discrete LEDs that can be sequentially energized to provide the desired readout

Nonvisible infrared light emitting diodes are used in conjunction with other components to provide a variety of applications, including:

- Signal detecting systems

- A wide range of opto-couplers (also called opto-isolators)

- Opto-coupled solid-state relays (SSR)

- Fiber optic communications systems

In many applications, moderately accurate visual indications of temperature, barometric pressure, speed of a moving vehicle and other information may only be required. The electrical analog of these changes, in linear voltage form, can be used to drive either a linear or a circular array of LED lamps for very effective measurements. Examples of these types of displays are shown in Figure 101.

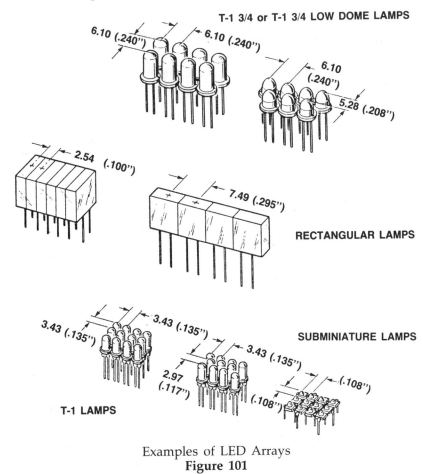

Examples of LED Arrays
**Figure 101**

Compared with analog meters, LED linear and circular arrays provide:
● Higher reliability at lower cost
● Greater resistance to mechanical shock and vibration
● Higher visibility in low and moderate ambient lighting
● A fast means of interpreting monitored information, making possible, in some circumstances, quicker decisions based on the different LED colors

Other uses for discrete light emitting diodes include those applications which provide a light source to be used for signal detection by electromechanical light sensing mechanisms and not by human vision. These signal detecting applications include:

- End-of-tape or end-of-ribbon sensing devices used in tape recorders, printers and other systems using similar techniques

- Punched card reading mechanisms

- Assembly line parts-counting and monitor mechanisms

- Tape loop stabilizing (max and min loop sensors) systems used in tape recording equipment

- Optical tachometer for motor speed control

- Bar code scanning mechanism for the Universal Product Code (UPC) in Point of Sale (POS) equipment

- Carriage travel sensing mechanism in printers, plotters and copying machines

- Smoke detector systems

- Densitometer mechanism for chemical analysis

- Liquid level monitor mechanism for both clear and opaque liquids

- Interlock circuits used on enclosures to provide safety in high-power electrical generating and radio and TV transmitting equipment

# REINFORCEMENT EXERCISES

Answer TRUE or FALSE to each of the following statements:

1. The specific color of light emitted by a LED is dependent on the substrate material and the concentration of dopants used to produce the P-type and N-type regions. Red, yellow and orange are produced by using gallium arsenide phosphide. Green is obtained from gallium phosphide and infrared from gallium arsenide.

2. A light emitting diode (LED) requires a reverse bias voltage (positive to the cathode) to cause light to be emitted.

3. The color of the light emitted from an LED could be red, orange, yellow, green or nonvisible infrared.

4. Removal of the voltage applied to the LED, or applying a reverse bias, will produce zero current in the diode resulting in no light emission.

5. Light emitting diodes, using relatively little power, are suitable for battery-powered equipment. They are shock and vibration proof, can operate within the temperature range from -55°C to +100°C and since they are semiconductors, have all the reliability features inherent in this technology.

6. Increasing the current flowing in the LED will result in a change of color being emitted by the device, in addition to an increase in light intensity.

7. The color of the emitted light can be changed by changing the color of the epoxy that encapsulates the LED.

8. The peak inverse voltage rating (PIV) of a light emitting diode is about the same as a silicon diode.

9. Because the estimated life of an LED is about 1 million hours (114 years), there should be no reason to replace the device if it is designed into the equipment properly and is operated within its rated limits.

10. Knowing the peak wavelength (color) and luminous intensity (amount of radiated light) of a light emitting diode will

provide enough information to sufficiently specify the optical characteristics of the device.

11. Luminous intensity is a function of LED current. As the current increases, the light emitted from the LED increases linearly.

12. Visible light emitting diodes are used as single light indicators in simple visual indicator circuits and as indicator arrays with multiple LEDs that are energized in a sequential manner to provide a custom readout.

13. No light emitting diodes will have good readability when exposed to the direct rays of the sun.

14. Analog meters are more rugged and less sensitive to shock and vibration than a light emitting diode array.

15. Since the supply voltage used with a light emitting diode is normally very low (5 to 6 volts), it is not necessary to use a series resistor when applying voltage to the LED.

16. Some LED devices incorporate a series chip resistor inside the LED package eliminating the need for an additional resistor.

17. A precision meter is always preferred as a visual indication of some parameter (e.g., temperature, barometric pressure, motor speed) rather than a light emitting diode readout.

Answers to the reinforcement exercises are on page 323.

 CHAPTER
ELEVEN

# LED DISPLAYS

# LED DISPLAYS

A light emitting diode display consists of a group of LEDs that generate light when energized. They are structured as a cluster of segments or as a dot-matrix array providing many variations of numerals, letters and/or special symbols for digital readouts. Analog displays are available in the form of multi-segment bar graphs. LED displays are produced in four colors – red, yellow, orange and green in a variety of sizes, packages and fonts.

A *font* is an arrangement of display elements in an individual letter or numeral (alphanumeric) and determines the amount of information that needs to be shown and the complexity of the associated circuitry. LED digital displays include 7, 9, 14, 16-segment fonts and several dot-matrix configurations.

Segmented displays have simpler associated circuitry compared with dot-matrix displays. Although they are more commonly used, they have limited display capability. Single and multiple-digit displays include:

● Carry-over numeral of "1", or the *overflow* display, e.g.:
  In adding $9 + 3$, the answer is 12. The "2" is in the furthermost right position and the carry-over numeral is the "1".
● Special characters, such as decimal point (.), colon (:) and the $2-$ segment polarity indication ($+$ or $-$).

Segmented bar graph displays have no standard number of segments, using the number of segments that are necessary for a specific application.

Dot-matrix digital displays have unlimited alphanumeric and graphics capability but require complex and extensive encoding and decoding circuitry.

## SEGMENTED DIGITAL DISPLAY

The 7-segment digital display (Figure 102), is the simplest, least expensive and is the most commonly used. The appropriate segments are energized to provide the desired alphanumerics. It can provide all ten numerals (0 to 9), but only nine upper case letters of the English alphabet: A, C, E, F, H, J, L, P and U.

Seven-Segment LED Digital Display
**Figure 102**

Despite its limitations, this display is applicable for use in digital watches, meters, thermometers, speedometers, odometers and similar digital readout applications.

NINE-SEGMENT DIGITAL DISPLAY
The *hexidecimal* (base 16) numbering system, inherent in most computers, uses 16 characters – the numbers "0" to "9" and the letters "A" to "F". To obtain upper case letters, "B" and "D" (not available in the standard 7-segment digital display), two additional segments must be provided. See Figure 103. The "B" can now be distinguished from the "8" and the "D" from the "0". This display is often referred to as a 9-segment display.

Nine-Segment Digital Display for Hexidecimal Readout
**Figure 103**

For readouts requiring the entire English alphabet, punctuation marks and several special symbols, 14-segment and 16-segment digital displays are available. Examples of these displays are shown in Figures 104 and 105.

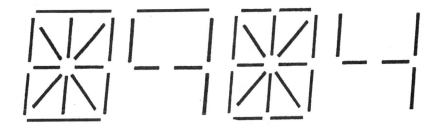

14-Segment Digital Display
**Figure 104**

16-Segment Digital Display
**Figure 105**

In addition to single-digit displays, multiple-digit display clusters are common. For example, an eight-digit, 7-segment display is used as the readout in a hand-held calculator. For use in desktop calculators, 10-digit and 12-digit displays are available. As with the single-digit displays, multiple-digit display clusters include a variety of packaging: the single-in-line package (SIP), the dual-in-line package (DIP) and special assemblies. The largest LED display presently available is 1 inch in height and is readable at a distance of 70 feet.

Examples of LED Displays
**Figure 106**

THE OVERFLOW DIGIT
The "overflow" digit "1" is used in digital meters, 12-hour clocks, watches and similar "carry-over" displays. It is available as a single-digit or as part of a multiple-digit display. In a multiple-digit display, the overflow digit is generally specified as a "½ digit". For

use in digital meters, an additional 2-segment section is added to the ½ digit to be used for polarity indication. Depending on the precision required for the digital readout, different multiple-digit displays are available. A 3½ digit display is shown in Figure 107.

For a 12-hour clock display, a colon (:) is used between the 1½ digit "hour" and the 2-digit "minute" display and is called a "3½ digit display with colon". See Figure 108.

3½-digit Digital Meter Display        3½-digit 12-hour Clock Display
**Figure 107**                                    **Figure 108**

A multi-range digital meter has a decimal point for each digit and can be activated as required for a specific voltage range.

## SEGMENTED BAR GRAPH DISPLAY

To convert analog information into a visual display, bar graph displays are used effectively in place of the more traditional analog panel meters. These displays are available in a wide variety of sizes, numbers of segments and packages. A typical 10-segment bar graph display in a 20-pin dual-in-line package is shown in Figure 109. Additional end-stackable packages are used for the continuous extension of the bar graph display.

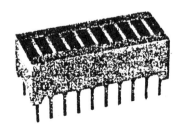

10-Segment Bar Graph Display in DIP
**Figure 109**

## DOT-MATRIX DIGITAL DISPLAY

To overcome the limitations of a segmented digital display, LED displays are also produced in dot-matrix form. The 5 x 7 matrix (5 columns of 7 rows) can display the 64 character ASCII set that includes all ten numerals (0 to 9), all 26 upper and lower case letters of the English alphabet, common punctuation marks and special symbols. The term ASCII is an acronym for American Standard Code for Information Interchange.

As with the segmented digital displays, dot-matrix displays are available as single-digit and multiple-digit displays in a variety of packages, matrices and sizes. A typical dot-matrix display, with left-hand decimal, is shown in Figure 110. The diodes at the intersecting rows and columns are energized to illuminate the required elements of the matrix.

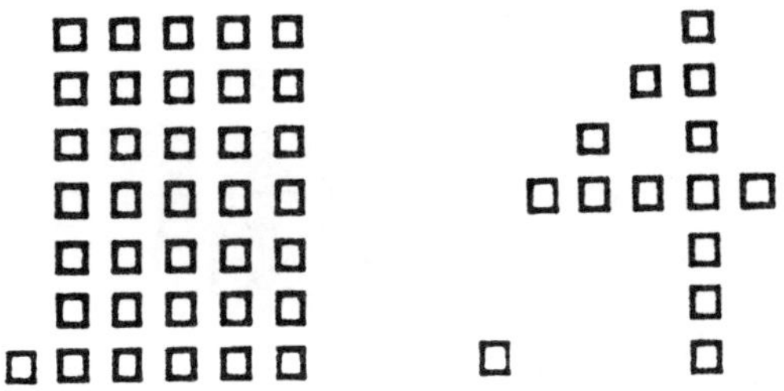

5 × 7 Dot-matrix LED Digital Display with Left-hand Decimal
**Figure 110**

A modified 4 × 7 dot-matrix display, used primarily to display only numerals, is shown in Figure 111.

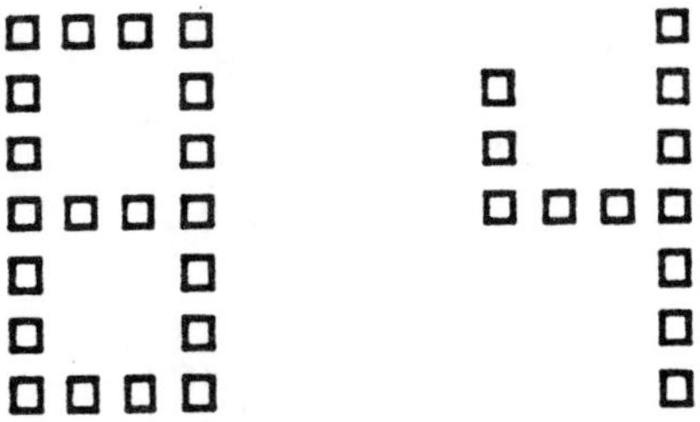

Modified 4 × 7 Dot-matrix LED Digital Display
**Figure 111**

Other matrices are available for larger-sized LED digital displays to provide a more pleasing appearance.

# DISPLAY MODES

### DIRECT-VIEW DISPLAY
In this display, the energized LED is viewed directly, a technique that was used in the original generation of LED displays. For a ¼ inch segmented digital display, 16 discrete LED chips, assembled in a straight line and connected in parallel, formed a single segment. Larger digital displays used long rectangular chips for each segment.

### REFLECTOR DISPLAY
Newer LED displays incorporate a reflector within each segment to gather the emitted light and distribute it evenly across the segment. See Figure 112. The light is reflected onto the viewing surface after being emitted by a small LED chip. This provides an evenly illuminated segment appearance at reduced cost.

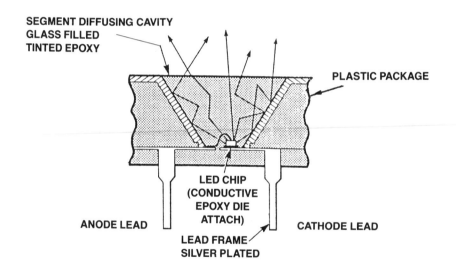

Reflector Display
**Figure 112**

### LIGHT-PIPE DISPLAY
For all practical purposes, the light-pipe display produces the same results as the reflector display. A light-pipe is a transparent rod that transmits light from one end of the rod to the other, even when bent. Light is gathered from a small LED chip buried at the base of the pipe and diffused across its viewing surface. This method reduces cost and improves the appearance of the display over first generation designs. See Figure 113.

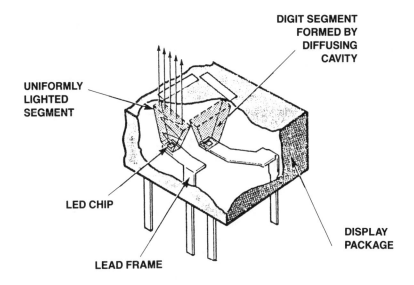

DIGIT SEGMENT
FORMED BY
DIFFUSING
CAVITY

UNIFORMLY
LIGHTED
SEGMENT

LED CHIP

DISPLAY
PACKAGE

LEAD FRAME

Light-Pipe Display
**Figure 113**

MAGNIFIED DISPLAY
For some applications, very small
LED displays can be constructed
for assembly into small areas. The
actual display is made as small as
possible with the image magnified
to appear as large as 0.125 inches.
These displays are used in digital
thermometers and digital meter
probes, but are available only as a
multi-digit, 7-segment display. A
typical package is shown in Figure
114, however, other assemblies are
available.

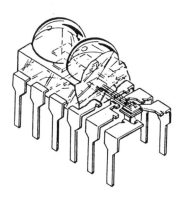

Magnified Display
**Figure 114**

All LED displays, regardless of the type of font, are available
either in a *common anode* or *common cathode* configuration. The segments or dots are energized as follows:

- COMMON ANODE – With all the anodes of the LED segments
  or dots connected to a common terminal and a positive D.C.
  voltage applied to this terminal, connecting any cathode terminal through a resistor to the negative terminal of the voltage

supply will energize that segment or dot. See Figure 115. Each set of contacts represents an electronic switch that is closed, when required, by its associated logic circuit, causing the desired LED segment of the display to be energized.

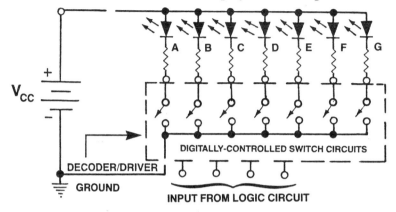

Energizing Segment or Dot of a Common Anode LED Display
**Figure 115**

- COMMON CATHODE – With all cathodes of the LED segments connected to a common terminal and with this terminal connected to the negative terminal of the voltage supply, applying a positive D.C. voltage through a resistor to any anode terminal will energize that segment. See Figure 116. Each set of contacts represents an electronic switch that is closed, when required, by its associated logic circuit, causing the desired LED segment of the digital display to be energized.

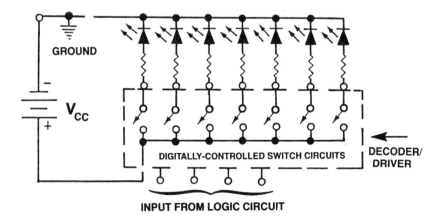

Energizing Segment or Dot of a Common Cathode LED Display
**Figure 116**

# DISPLAY DRIVING TECHNIQUES

There are two methods of driving a display:

● Direct D.C. drive

● Strobing or multiplexing

### DIRECT D.C.DRIVE
Each segment of a digital display, or each segment of a bar graph display, is continually illuminated. Each single digit, or segment of a bar graph display, has its own *decoder/driver* circuit.

The block diagram in Figure 117 shows the associated circuitry required to convert the output digital code of a digital logic system into voltages that will energize the appropriate segments of a 7-segment display to provide readable information.

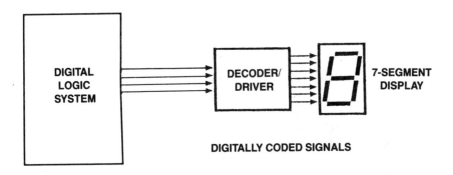

Decode/Driver and Display Circuitry
**Figure 117**

### DECODER SECTION
The *decoder* section of the circuit converts a digitally coded signal into the appropriate voltages to energize the desired segments. The conversion and energized segments create a display of the decimal equivalent of that selected character.

### DRIVER SECTION
The display element *drivers* act as the interface between each decoder output and the corresponding display elements to amplify the voltage and/or current level required to drive those display elements.

The advantages of the direct drive approach are:

• Drive circuitry is simpler than that used for multiplexing.

• The decoder section does not have to handle the relatively high current levels needed to energize the LED segments.

The disadvantage of the direct drive appoach is that the LEDs are less efficient (higher power is being dissipated in the LEDs) when D.C. driven than when multiplexed for the same light intensity that is being emitted.

MULTIPLEXED OR STROBED DRIVE
Multiplexing, or strobing, is a drive technique that can be used with multiple-digit displays (generally more than 4) to save power and system cost.

LED displays operate at increased efficiency when they are pulsed (strobed) at a high peak current and at a low duty cycle resulting in reduced average power dissipation when compared to a direct D.C. drive technique.

In addition, some of the complex drive circuitry is shared by all the digits. The total drive power required is reduced and fewer parts are used, lowering overall system costs.

In the circuit of Figure 118, six 7-segment digital displays are used in an example of a multiplexing system.

• At the first clock pulse, the Multiplexing Logic Circuit selects Digit Data Storage Buffer #1. This digitally coded data is applied to the input of the single Decode/Driver circuit and the corresponding segments for all digits are energized. The Digit Scanner only enables Digit Driver #1 and only the first digit is displayed.

• At the second clock pulse, the Multiplexing Logic Circuit selects Digit Data Storage Buffer #2 and the second digit data is applied to the Decode/Driver. Once again, the corresponding segments are energized. Since the Digit Scanner only enables Digit Driver #2, only the second digit is displayed. Each digit is enabled by its own digit driver.

• This routine continues through Digit Drivers #3 to #6. The procedure is repeated as the multiplexing process continues.

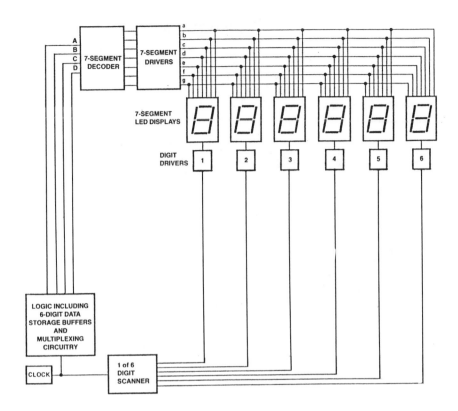

Multiplexed 6-digit LED Digital Display
**Figure 118**

Other LED displays (bar graph and dot-matrix) can be multiplexed in essentially the same manner. As the number of elements making up the display, and as the number of different characters required to be displayed increases, the complexity of the multiplexing system will also increase.

This drive technique is possible because the human eye cannot detect that the LED segments are actually turning OFF even though the short duty cycle pulse has been removed. When the next high peak current pulse is applied to the segment, the eye continues to perceive the bright intensity display as a steady-state condition and not as a pulsed or strobed burst of light.

To produce a flicker-free display, the strobing rate should be a minimum of 100 Hz, however, a rate of 1 Khz or higher will optimize the overall performance of a 7-segment display.

# ONBOARD INTEGRATED CIRCUITS

Both single-digit and multiple-digit displays are available with decode/driver circuits provided as an integrated circuit and assembled on the same substrate as the display. It is referred to as an *onboard integrated circuit* (OBIC) display. An example is shown in Figure 119.

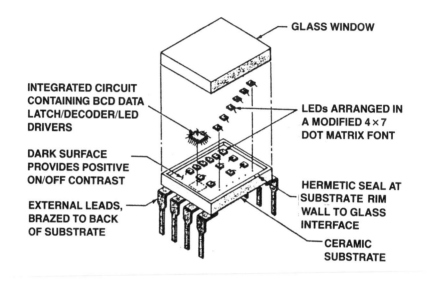

Display with Onboard IC Circuit in Same Package
**Figure 119**

This approach is more expensive than one using a display that does not incorporate an OBIC, however, the following advantages can offset the higher initial cost:

● Number of pin connections on the display package is reduced

● Handling and assembly costs are reduced

● Display system design is simplified

● PC board space is saved

● Overall system reliability is improved

The disadvantage of an OBIC display is that as more functions are added, the display package becomes unique and a second source of supply may become difficult to obtain.

# DISPLAY SELECTION CRITERIA

Display selection combines objective technical considerations with subjective human esthetic values. In all cases, however, the selection of a specific display depends on its end use. The following factors contribute to the final selection:

- VIEWING DISTANCE AND AMBIENT LIGHT – Largely determine the size and brightness of the display.

- POWER REQUIREMENTS – Includes the power dissipated by the display and its associated electronic circuitry.

- COST – Total cost must include the display cost and all associated circuitry, power supply and hardware.

- HUMAN FACTORS – Although difficult to evaluate quantitatively, these factors are probably the most important considerations in the usability and success of a display. Human subjectivity determines the selection of size, shape and color.

## LED DISPLAY SPECIFICATIONS

### PRODUCT DESCRIPTION AND DEVICE FEATURES

This information is on the data sheet and includes the number of digits (or segments), color, font, size and capabilities of the display. In addition, the characteristics and applications of the display are given, as well as the recommended value of the D.C. supply voltage.

### PACKAGE OUTLINE

The drawing of the package appears on the data sheet and includes information on package dimensions, mounting details, lead designation, lead material, lens color and type (diffused or clear).

## ABSOLUTE MAXIMUM RATINGS
### (At 25°C case temperature)

PEAK REVERSE VOLTAGE – Per segment (PIV)
Maximum reverse voltage (negative voltage to the anode) that can safely be used and still have the segment (diode) maintain its non-conductive state. Typical values range between 3 to 6 volts. Although this rating is shown on the data sheet, and is much lower than the PIV of silicon, the LED segment (diode) is generally turned OFF by reducing the voltage to zero.

CONTINUOUS D.C. FORWARD CURRENT ($I_{Fcont}$)
Maximum allowable current for continuous D.C. operation.

- Per segment (or diode) – Values from 10 to 40 milliamperes
- Total display – Values from 80 to 320 milliamperes

PEAK FORWARD CURRENT ($I_{Fpeak}$)
The maximum allowable peak forward current for non-continuous operation. The rating is based on short duration pulse widths, typically less than 100 microseconds at 300 pulses per second (less than a 3% duty cycle). Different pulse widths and repetition rates can be used. Typical peak forward current values range from 100 to 400 milliamperes per segment.

D.C. POWER DISSIPATION – Per digit ($P_D$)
The maximum power capability depends on the size of the LED chip(s), the type and size of the package and its mounting features. D.C. power dissipation values range from 150 milliwatts to 1.5 watts. The power capability of the device must be derated as the case temperature increases. In applications where the power dissipation causes the case temperature to rise excessively, heat sinking is required with adequate cooling to keep the case temperature within its rated limit.

If the logic circuits are included in the same package, the total power capability of the device is specified.

OPERATING AND STORAGE TEMPERATURE

- Plastic devices: from −20°C to 85°C. This range can vary among different types of displays, i.e.: some devices range from −55°C to 100°C.

- Hermetically sealed devices: range from −55°C to +90°C or from −55 to +100°C.

ELECTRICAL AND OPTICAL CHARACTERISTICS
(At a case temperature of 25°C, unless otherwise noted)

FORWARD VOLTAGE ($V_F$)
This voltage is the sum of the threshold voltage, $V_{TH}$, and the voltage across the resistance, $R_F$, of the LED. See CHAPTER THREE – DIODE CHARACTERISTICS AND SPECIFICATIONS for a discussion of $V_{TH}$ and $R_F$.

For light emitting diodes, $V_F$, is a function of the substrate and dopant selected for the desired wavelength (color) emitted by that display. Typically, these values range from 1.7 volts to 3.0 volts with a temperature coefficient (TC) of approximately -1.5mv/°C.

REVERSE CURRENT – Per segment ($I_R$)
This is the maximum reverse current measured at the peak reverse voltage value. Typical values are from 5 nanoamperes to 100 microamperes at 25°C ambient temperature depending on the specific device.

CAPACITANCE – Per segment
The value of PN junction capacitance is measured at zero voltage. Typical values range from 25 to 700 picofarads depending on the geometry of the chip and package size.

SPEED OF RESPONSE
The time required to turn the device ON and OFF. See CHAPTER THREE – DIODE CHARACTERISTICS AND SPECIFICATIONS for a detailed discussion of this characteristic. Typical speed of response values for LED displays range from 20 to 100 nanoseconds and is most suitable for multiplexing drive applications.

LOGIC CIRCUIT CHARACTERISTICS
For displays containing an onboard integrated circuit, the electrical characteristics of the latching and decode/driver circuits are listed in the data sheet. These include typical and maximum values of supply current for the logic circuits at a specified supply voltage, and typical and maximum values of logic input currents for its ON and OFF states.

PEAK WAVELENGTH (COLOR)
This characteristic defines the color of the LED display and is specified in nanometers or Angstrom units. It is given as a typical and maximum value and is measured at the peak of the radiated spectral response curve. To enhance contrast in a high ambient light

environment, optical filters with the same color as the emitted light are used. Peak wavelength is used to determine the percentage of emitted light passing through the filter. If a filter has a relative transmission of 90% at a given peak wavelength and 20% at off-peak wavelengths, then 90% of the desired color at peak wavelength will pass through the filter while 80% of the off-peak wavelength light, such as reflections from outside the display, will be absorbed.

LUMINOUS INTENSITY – Per segment (I)
The intensity of the emitted light is specified at some value of forward current ($I_F$), in candela (c), millicandela (mcd), or microcandela (fcd). The *lumen*, another unit of measurement for luminous intensity at a specified wavelength, is sometimes used. One candela is equal to 12.56 lumens. Luminous intensity is a measurement of the number of lines of light (luminous flux) that radiate from a light source in the form of a cone. This cone of light radiates perpendicularly from the face of the display surface that contains the source of light.

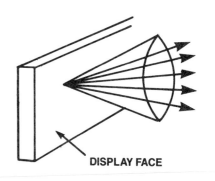

DISPLAY FACE

Luminous Intensity
**Figure 120**

Of all the colors in the light spectrum, the human eye is most responsive to yellow-green. Any changes in luminous intensity are detectable by the unaided human eye only when the differences exceed 40%. For properly matched segments, the value of luminous intensity can not deviate more than plus or minus 40% between segments of a display. In a matched multiple-digit display, this same approach – limiting the maximum deviation of plus or minus 40% in luminous intensity between digits is maintained.

# REINFORCEMENT EXERCISES

Answer TRUE or FALSE to each of the following statements:

1. Light emitting diode displays can be in the form of a segmented digital display, a segmented linear (analog) display
or a dot-matrix digital display.

2. The LED 7-segment digital display has limited capabilities; it
can only display the numbers 0 to 9 and nine upper case letters (A,C,E,F,H,J,L and P). Because of these limitations, it is
only used for special display applications.

3. The dot-matrix digital displays have unlimited alphanumeric
and graphics capabilities but require extensive decoding and
encoding circuitry associated with the display.

4. A 16-character hexidecimal readout requires either a 14-seg-
ment or 16-segment digital display.

5. The multiple-digit LED display used for a 12-hour clock or
watch is called a 4-digit display with colon.

6. There are four types of LED display modes that include:
direct-view, reflector, light-pipe and magnified displays.

7. The reflector and light-pipe displays produce the same result
– they provide an evenly illuminated segment appearance at
the viewing surface while still using a single low cost LED
chip as the source of light.

8. Multiplexing is the preferred drive technique over a direct
D.C. drive in a single-digit LED display.

9. In a 4-digit (or more) LED display, the direct D.C. drive technique used for displays use less power than a multiplexing
(strobed) technique. The direct D.C. drive technique has
fewer parts in the circuit and offers lower overall costs.

10. Multiplexing is accomplished by pulsing (strobing) the digits
with a controlled, short pulse-width (low duty cycle), high-
current pulse. This drive technique produces a bright display
that will appear ON to the human eye even when the
segments are in their pulsed OFF state. This technique will

provide relatively low power dissipation in the display section.

11. Flicker-free operation of the LED multiplexed display can be achieved at a rate below 60 Hz.

12. Because of its higher initial cost, the decode/driver and latch logic circuits used in conjunction with LED displays are never enclosed in the same package as the display.

13. In specifying a particular LED display, human factors are only minor considerations in the final selection criteria.

14. The speed of response of LED displays range between 20 to 200 nanoseconds, making these displays suitable for use with multiplexing drive techniques.

15. The use of special filters in conjunction with LED displays can enhance display contrast and optimize readability in high ambient light environments.

16. Of all the colors in the light spectrum, the human eye is most responsive to red.

17. Differences in light intensity are not detectable by the unaided human eye when the deviation is less than 40%.


Answers to reinforcement exercises are on page 325.

 CHAPTER
TWELVE

# OTHER DISPLAY TECHNOLOGIES

### GAS DISCHARGE DISPLAYS
### VACUUM FLUORESCENT DISPLAYS
### LIQUID-CRYSTAL DISPLAYS
### REINFORCEMENT EXERCISES

# OTHER DISPLAY TECHNOLOGIES

## GAS DISCHARGE DISPLAYS

Gas discharge displays are indicators of alphanumeric and miscellaneous graphic information, using the ionization of neon gas as the basis of their operation. As current flows in the device, an orange glow is emitted from the gas and provides the desired readout.

### GAS DISCHARGE DISPLAY CONSTRUCTION

A typical gas discharge segmented display structure is shown in Figure 121. Contacts, connections and segments of the multiple-digit display elements are silk-screened with a conductive material. The display segments and its connected contacts are designated as the cathode and the other set of contacts as the anode. The plates are separated by a spacer, joined, filled with a neon gas mixture and then sealed.

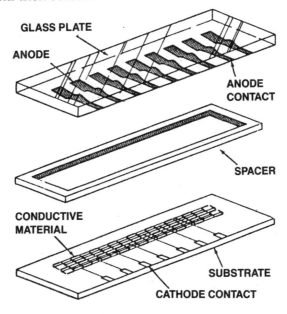

Typical 7-segment Gas Discharge Display Structure.
**Figure 121**

Because of the simplicity of the screen printing process that is used, any desired font or pattern can be easily screened on the glass. This includes bar graphs, in any desired form or size, both linear and circular, text of any nature or language and special symbols or graphics. To maintain a good hermetical seal, however, the maximum size of the glass plates are usually held to about 4 inches by 4 inches, limiting the height of the display to about 3 inches by 3 inches.

Where good readability in high ambient light and/or viewing from long distances are required, the gas discharge display would be suitable. This type of display is called a *planar* or *plasma* gas discharge, or *cold cathode* display. It is used most frequently in large digital clocks, large segmented analog readouts (bar graphs) and in desktop calculators operating in a highly lit room.

Another gas discharge display structure in dot-matrix form is shown in Figure 122.

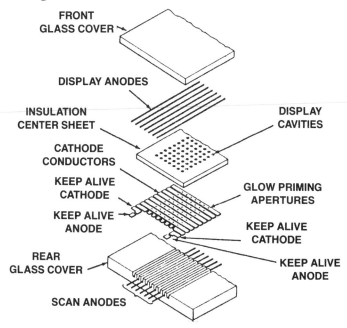

Typical Dot-matrix Gas Discharge Display Structure
**Figure 122**

This display consists of individual neon gas-filled cavities surrounded by glass plates plus conductive dot-matrix elements inside a sealed envelope. Anode and cathode electrodes at the display elements are connected to the external terminals.

## PRINCIPLE OF OPERATION

A gas discharge display operates by the principle of emission of light from a gas caused by current flowing through the gas. To produce current flow, the *ionization potential* or *break down voltage* of the neon gas must be exceeded, reducing its resistance and allowing current to flow.

### SEVEN-SEGMENT DISPLAY
When the gas between the anode segment of a display and the cathode contact is ionized, current can flow through this individual section that emits its characteristic orange glow to provide a readable display.

### DOT-MATRIX DISPLAY
Any cell of the display can be lighted by applying a high D.C. voltage across the electrodes addressing that location. The high voltage will ionize the neon gas, causing current to flow through the selected cell of the display. This produces a bright orange glow visible through the front glass layer.

To ionize the neon gas, a high D.C. supply voltage (from 150 to 250 volts, depending on character size) is required. Once the neon gas is ionized, current flow through the segment, or cell, is sustained at a voltage significantly lower than the ionizing voltage (about 20 to 30 volts lower). The voltage differential between the supply and sustaining voltage is absorbed by the resistor in series with the supply voltage. See Figure 123.

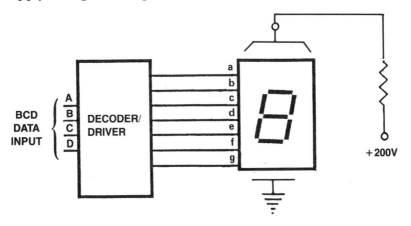

Gas Discharge Display with Required Interface Circuitry
**Figure 123**

The current level is controlled by the series resistor. The greater the current, the brighter the display. Typically, gas discharge diplays operate at a current level of 100 to 200 microamperes per segment.

A green display can be produced by coating the front glass with an appropriate phosphor to absorb the ultraviolet energy being emitted by the neon gas, thereby releasing secondary light photons in the green region of the visible light spectrum. With the use of a filter, a red or amber display is possible.

The interface circuitry that is required with a gas discharge display (Figure 123) operates essentially in the same manner as discussed in the previous chapter, although electrical parameters will differ from those that apply to LED displays. A latch circuit is used to maintain digital information coming from a digital system. A decode and driver circuit, with appropriate logic level voltages, converts coded digital information into the high voltages needed to energize the appropriate display segment or cells.

These displays are designed to operate in either the direct D.C. drive or multiplexing/ mode.

## GAS DISCHARGE DISPLAY CHARACTERISTICS

Compared with LED displays, the following must be considered:

ADVANTAGES:
- Gas discharge displays are more applicable to high ambient light environments than LED displays.
- Since gas discharge displays can be as tall as 3 inches, they are visible from a distance of over 150 feet.
- Using a screen printing process, any shape or form of display is easily made (bar graphs, linear, circular, etc.).

DISADVANTAGES:
- Since a high D.C. voltage is required for energizing of a gas discharge display, it is not readily suitable for use in equipment using a low voltage (5 to 12 volts) power supply.
- A gas discharge display operates over a limited temperature range of 0°C to + 55°C and is not shock and vibration proof.
- The estimated life span of a gas discharge display is about 100,000 hours (12 years), one-tenth that of the LED display.

## SPECIFICATIONS – GAS DISCHARGE DISPLAYS

### PRODUCT DESCRIPTION AND DEVICE FEATURES

The data sheet presents information on the number of segments, digits, color, font style, size and capabilities of the display. In addition, the characteristics and applications of the display are given, as well as the recommended value of the D.C. supply voltage.

### PACKAGE OUTLINE

The drawing of the package appears on the data sheet and includes information on package dimensions, character height and width, mounting details, terminal spacing, terminal designations and lead material.

### ABSOLUTE MAXIMUM RATINGS
(At 25°C ambient temperature)

DISPLAY ANODE VOLTAGE ($V_B{}^+$)
The maximum allowable D.C. supply voltage between the display anode terminal and the common reference terminal. Typical values are from 150 to 300 volts.

DISPLAY CATHODE VOLTAGE ($V_K$)
Voltage is negative with respect to the reference. Prior to ionization, voltage between the common anode and any cathode may equal this voltage and no damage will occur to the display. Display voltage can range between -100 and -250 volts. Peak cathode current must be limited to the absolute maximum rating.

PEAK CATHODE CURRENT – Per segment ($I_K$)
A value that depends on the number of digits in the display. A typical value for a 16-digit display is 600 microamps.

TEMPERATURE RANGE
Operating temperature – Typically from 0°C to +55°C
Storage temperature – Typically from -40°C to +70°C

OTHER ENVIRONMENTAL RATINGS
These indicate the capabilities of the display for altitude, relative humidity, shock, vibration and life expectancy.

ELECTRICAL AND OPTICAL CHARACTERISTICS
(At an ambient temperature of 25°C, unless otherwise noted)

PANEL VOLTAGE DROP
The voltage between anode and cathode with the display ON having typical values between 140 to 160 volts.

SEGMENT UNIFORMITY
With a specified cathode current, the segment is considered to have a uniform intensity when the cathode glow appears uniform to the unaided eye while being viewed from a distance that is more than 12 inches. Under normal operating conditions, the only glow comes from an energized digit.

LUMINOUS INTENSITY
A value that is measured with a calibrated photometer (light measuring meter) that is mounted so that the sensing head of the photometer is parallel with the surface of an unfiltered display and focused on the center of a segment under test when operated at its specified conditions.

For optimum operation of the display, a suitable nonreflective matte filter is supplied with the display package to enhance contrast ratio.

See CHAPTER ELEVEN – LED DISPLAYS for a detailed discussion of this parameter.

HORIZONTAL AND VERTICAL VIEWING ANGLES
Minimum values specified with the display mounted in a vertical position. Typical viewing angle values:
Horizontal = 125°
Vertical = 115°

Electrical characteristics of the latching and decode/driver circuits are generally specified in a data sheet. They include the typical and maximum values of supply current for the logic circuits at a specified supply voltage and typical and maximum values of logic input currents for its ON and OFF states. Any additional information required for blanking, timing or other specific operations is included.

New plasma display dot-matrix modules are available in a wide variety of styles. They range in capability from 2 lines x 16 characters per line to 12 lines x 40 characters per line.

# VACUUM FLUORESCENT DISPLAYS

Vacuum fluorescent displays are essentially modified vacuum tubes with phosphor-coated anodes that emit light when bombarded with electrons emitted from their cathode. A display basically consists of a directly heated filament (cathode), a control grid and a phosphor-coated glass screen, or plate, that displays the readout.

Two types of vacuum fluorescent displays are available:

CATHODE RAY TUBE (CRT)
A device used as the picture tube in a television set or as a monitor in a computer or radar console. Known as *cathodluminescence*, this form of display is created by the collision of high energy electrons with a phosphor coating applied to the inside surface of the glass screen in a CRT.

INDICATOR PANEL DISPLAY
Vacuum fluorescent indicator panel displays have become the leading display technology for automotive use because of their compatibility with automotive electronics, their relatively low cost and general customer acceptance. It is felt by most automobile manufacturers that vacuum fluorescent indicator panel displays allow for larger letter and number images, are easy to see under adverse conditions and conform to traditional instrumentation panel lighting techniques.

## GENERAL PRINCIPLE OF OPERATION

- If an A.C. or D.C. voltage is applied to the filaments of a vacuum fluorescent tube, the filaments (cathode) will heat to an incandescent state and emit a flow of electrons in the direction of its anode that is being energized by a separate D.C. voltage. The anode is either the inner surface of the CRT phosphor-coated screen, or the inner surface of a glass viewing plate of a segmented/dot-matrix display on which a fluorescent or phosphorescent material has been deposited.

- When the electrons are emitted from the cathode to bombard the phosphor-coated screen or glass viewing plate, a light is emitted from the face of the screen at the point of impact and provides a coherent display. One of a variety of colors is available, depending on the phosphor or filter.

## CONSTRUCTION AND OPERATION
## OF THE CATHODE RAY TUBE

The CRT consists of a large vacuum glass or ceramic envelope, with a neck at one end and a flared or cone-shaped section at the other. Inside the neck is a metallic structure, called the *electron gun*, containing a filament (cathode) in combination with a control grid and accelerating anodes.

In addition, there are a series of horizontal and vertical deflection plates used to direct the flow of electrons emitted from the cathode. The flared or cone-shaped section has a flat glass face or screen whose inner surface is coated with a fluorescent or phosphorescent material. See Figure 124.

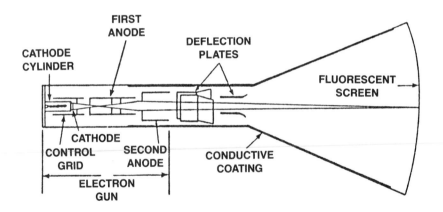

Cathode Ray Tube Structure
**Figure 124**

When a voltage is applied across the terminals of the filament, the heated filament or cathode will emit a stream of electrons in the direction of the screen. The magnitude of a separate voltage applied to the control grid determines the amount of electrons which impact the coated screen and controls the brightness of the image being emitted from the screen.

The accelerating anodes act to provide the high velocity for the electrons to energize the phosphor-coated screen areas that are being selected by the action of the horizontal and vertical deflecting plates. Depending on the phosphorescent coating, the emitted image is either a green, bluish-green, amber (yellow), blue, red or white light (all the combined colors).

## CONSTRUCTION AND OPERATION OF FLUORESCENT INDICATOR PANEL

Using thick film technology, a phosphorous coating is deposited on the front inside glass or ceramic plate surface (anode) of a segmented or dot-matrix display of any desired size and shape. A fine wire mesh (control grid) is placed behind each character or symbol. A filamentary structure (cathode) is mounted behind the grid structures. See Figure 125.

The entire assembly is then sealed in a glass-enclosed vacuum envelope with all the elements connected to external terminals.

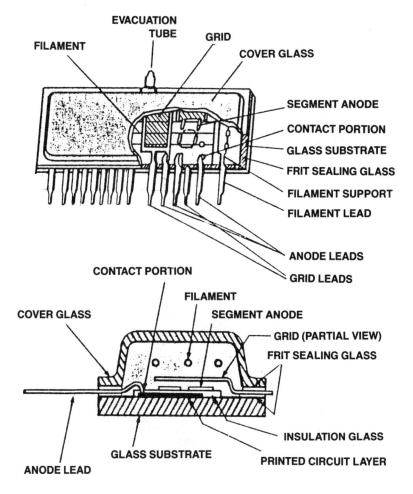

Fluorescent Segmented Indicator Panel Structure
**Figure 125**

When an A.C. or D.C. voltage is applied to the filaments, the heated cathode will emit a stream of electrons. When a separate D.C. voltage is applied to selected anodes and grids, the electron flow will be directed through the control grid to the appropriate segments or dots, causing the phosphorescent anodes to luminesce and emit light. By varying the duty cycle of the D.C. voltage applied to the anodes and control grids, the light intensity of the display can be varied.

The most common display color is bluish-green, however, with use of appropriate filters, red, orange or yellow light can also be emitted.

Several typical vacuum fluorescent indicator panels are shown in Figure 126.

Examples of Typical Vacuum Fluorescent Indicator Panels.
**Figure 126**

## SPECIFICATIONS – VACUUM FLUORESCENT INDICATOR PANELS

### PRODUCT DESCRIPTION AND DEVICE FEATURES

The data sheet presents information that includes the number of digits, character format, color, size, special graphics and capabilities of the display. In addition, the characteristics and applications of the display are listed as well as the recommended A.C. and D.C. supply voltages and the mode of operation (direct drive or multiplex). Multiplexing is referred to as the dynamic mode of operation and a direct D.C. drive is called the static mode.

### PACKAGE OUTLINE

The drawing of the package appears on the data sheet and includes information on package dimensions, character height and width, mounting details, terminal spacing, terminal designations and lead material. These displays are made in a wide variety of sizes, number of digits, font styles and packages.

### ABSOLUTE MAXIMUM RATINGS
(At 25°C ambient temperature)

PEAK ANODE VOLTAGE ($E_b$)
The maximum allowable voltage that may be applied between the anode and cathode. Depending on the device, values can range between 12 to 60 volts.

PEAK GRID VOLTAGE ($E_c$)
The maximum allowable voltage that may be applied between the control grid and cathode. Depending on the device, values can range between 12 to 60 volts.

PEAK ANODE CURRENT – Per digit (Multiplexing mode) ($I_b$/dig)
Depending on the device, this value can range from 1.5 to 15 milliamperes. In some cases, the peak current per segment is specified.

D.C. ANODE CURRENT – Per digit (Direct drive mode) ($I_b$/dig)
Depending on the device, this value can range between 0.2 to 2 milliamperes.

OPERATING AND STORAGE TEMPERATURE RANGE
There is no standard temperature range, however, it typically lies between 0°C and +150°C.

## ELECTRICAL AND OPTICAL CHARACTERISTICS
### (At 25°C ambient temperature, unless otherwise noted)

FILAMENT CURRENT AND FILAMENT VOLTAGE – $(I_f)$ $(E_f)$
Filament current is listed at a specified filament voltage. Depending on the device, $I_f$ ranges between 20 to 500 ma.
$E_f$ (D.C.) ranges between 1.3 and 1.7 volts.
$E_f$ (A.C.) ranges between 1.2 and 9.7 volts.

ANODE CURRENT – Per digit $(i_b/dig)$
This characteristic is listed as a maximum and typical value at a specified anode to cathode D.C. voltage and control grid to cathode voltage. Depending on the device, this value is sometimes specified as anode current per segment.

CONTROL GRID CURRENT – Per digit $(i_c/dig)$
This characteristic is listed as a maximum and typical value at a specified control grid to cathode voltage. Depending on the device, this value is sometimes specified as grid current per segment.

LUMINOUS INTENSITY (L)
A value measured with a calibrated photometer with its sensing element parallel to the surface of the indicator panel under test when operated at its specified conditions. See CHAPTER ELEVEN – LED DISPLAYS for a detailed discussion of this parameter.

BRIGHTNESS RATIO BETWEEN DIGITS – $L_{MIN}/L_{MAX}$
The relative difference between the minimum and maximum luminous intensity of the digits in the display. This value is expressed as a maximum percentage and is typically 50%.

HORIZONTAL AND VERTICAL VIEWING ANGLES
Minimum values specified with the display mounted vertically. Typical viewing angles: Horizontal $= 125°$; Vertical $= 115°$.

LOGIC CIRCUITRY CHARACTERISTICS
Vacuum fluorescent indicator panels are available with and without onboard integrated circuitry (OBIC) with the electrical characteristics of the latching and decode/driver circuits specified on the data sheet. They include typical and maximum values of current for the logic circuits at a specified supply voltage and typical and maximum values of logic input currents for its ON and OFF states. Any information required for blanking, timing or other specific operations is also included.

# LIQUID-CRYSTAL DISPLAYS (LCD)

Liquid-crystal displays (LCD) are extremely low-power, low-voltage readouts which operate on a principle of scattering light from an external source to provide a desired display. Depending on the device, the display is generally black or gray with other colors available by using different filters.

## LCD CHARACTERISTICS

ADVANTAGES:

● LCDs consume extremely low power, making them most suitable for use in battery-operated equipment, including digital watches, hand-held calculators, portable computer monitors, digital thermometers and portable digital meters.

● They are very readable in high-light intensity environments, particularly, direct sunlight.

● They operate at low voltage (1.5 to 25 volts) and can be driven by CMOS circuitry. LCD normal operating voltages are compatible with power supplies used in modern equipment.

● LCD font style and size can be easily customized for any unique application. Generally, the maximum character height is 4 inches, however, the size of the display can be increased by using special manufacturing techniques.

● With the proper use of filters, any display color can be obtained.

● Liquid crystal displays can be designed for either front lighted (reflective) or back lighted (transmissive) modes of operation.

● They are potentially the lowest-cost display on a per digit basis.


DISADVANTAGES:

● LCDs are not visible in the dark unless a source of light is available. A general approach is to include, as part of the display, a source of backlighting that can be switched ON or OFF as desired.

- LCDs operate over a limited temperature range – generally from 0°C to +55°C. Some of the newer LCD devices operate from −20°C to +70°C, with some special types designed to operate from −40°C to +85°C. The response time of the LCD slows as the ambient temperature drops and irreversible damage to the display will occur when frozen. Heaters and ovens are often used to raise the temperature to a higher value, however, the extra devices add to the system power requirements and cost.

- Generally, LCDs have a relatively narrow viewing angle, about 60° to 90°, but some newer displays provide a means of adjusting the viewing angle to overcome this problem.

- A.C. voltage must be used as the energizing voltage. If D.C. is used, defects in the cell will eventually result.

- LCDs are not shock or vibration proof and have an estimated life expectancy of about 100,000 hours (12 years), one tenth the life expectancy of LED displays.

## LCD CONSTRUCTION AND OPERATION

One feature of a liquid-crystal display is its simplicity of construction and ease of customizing. See Figure 127.

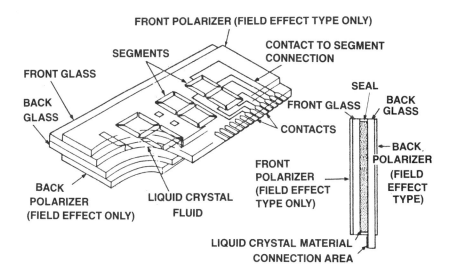

Construction of a Liquid-Crystal Display
**Figure 127**

An LCD structure consists of two plates of clear glass having thin layers of conductive, transparent material deposited on its inner surfaces. Indium oxide is generally used as the deposition material.

The uniformly deposited conductive area on the back glass plate acts as one electrode with a deposited conductor connected to a contact terminal. The desired display pattern is deposited on the front glass plate – a segmented display, dot-matrix display or custom layout, with connections made to the display pattern contacts.

A liquid-crystal material, called *nematic liquid*, is sandwiched between the two plates and the entire glass structure is then hermetically sealed at the edges. The contact area, or external terminal area, is extended outside the body of the display.

There are two basic types of liquid-crystal displays:

● *Twisted nematic field-effect* type

● *Dynamic scattering* type

The nematic field-effect type is the preferred choice for digital watches, hand-held calculators, digital meters and other portable instruments because of its low-voltage requirement (1.5 to 35 volts) and low current drain (typically 0.100 to 10 microamperes).

Dynamic scattering is generally selected for larger displays because it can be implemented more easily with larger packages and provides a wider, more uniform viewing angle.

It is important to note that, unlike all other displays, the LCD depends on ambient light to produce a visible display. This limitation is frequently overcome by the addition of an external light source (behind the display) that can be switched ON or OFF as desired.

## NEMATIC FIELD-EFFECT TYPE

In the nematic field-effect display, the top and bottom light control film (*polarizers*) are oriented so that their reflective surfaces are at right angles to each other. See Figure 128. When the LCD is not energized, light enters through the top of the display, goes through the untwisted LCD molecules and strikes the bottom polarizer where they are absorbed, providing no display.

When the LCD is energized by applying a voltage across the electrodes, the molecular structure of the liquid crystal is rotated 90°, twisting the light at the selected segments so that the bottom polarizer can now reflect the light back through the liquid crystal molecules to provide a display of the selected segments through the top plate.

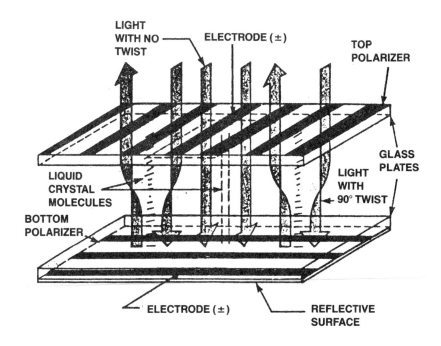

Nematic Field-effect Display Structure.
**Figure 128**

TRANSMISSIVE (BACK LIGHTED) TYPE (Figure 129a)
Depending on the orientation of the polarizer behind the back glass plate, the liquid crystal display can be made visible (transmissive) or opaque in its non-energized state. The two polarizers are oriented so that the front and back films are parallel, resulting in a display which has a light character on a dark background. See Figure 129a.

REFLECTIVE (FRONT LIGHTED) TYPE (Figure 129b)
The front and back polarizers are arranged at right angles to to each other. A diffuse reflector is attached to the back of the LCD structure and when energized, the display has a dark character on a light background.

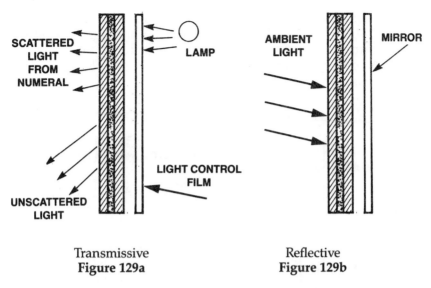

Transmissive
**Figure 129a**

Reflective
**Figure 129b**

Field-effect Transmissive and Reflective LCD Structures.

New LCD products are available for dual mode operation. They are called "transflective" and are made with a backing that reflects ambient light and transmits back light.

## DYNAMIC SCATTERING TYPE

When the dynamic scattering LCD is in a non-energized state, its molecular structure (lattice) is in a uniformly-oriented condition. If an A.C. (square wave) voltage is applied between the back electrode contact and the desired display pattern cell contact, the molecular order (lattice structure) is disrupted and the optical appearance of the desired display pattern is changed, providing a readable display.

This change is caused by the difference in the ambient light reflected, or scattered, by the energized display pattern compared with the light reflected by, or transmitted through the background field. The degree of optical change is called the *contrast ratio* and is the measurement used for liquid-crystal displays to provide a quantitative measure of the quality of the display. A typical LCD contrast ratio is 20:1.

TRANSMISSIVE (BACK LIGHTED) TYPE (Figure 130a)
The back light is directed through a light control film attached to

the back glass plate in a manner similar to that of a louvered Venetian blind that is blocking light from coming through. The light is only scattered for viewing in the area defined by the energized display pattern and produces a light character on a dark background.

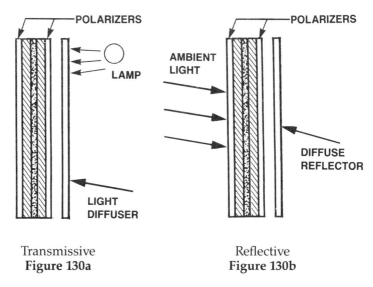

Transmissive
**Figure 130a**

Reflective
**Figure 130b**

Dynamic Scattering Transmissive and Reflective LCD Types

REFLECTIVE (FRONT LIGHTED) TYPE (Figure 130b)
The transmissive display can be easily converted into a reflective type by substituting a mirror for the light control film. The display is still operating on the principle of light scattering, but now only ambient light from the front of the display is used. As before, the display produces a light character on a dark background.

## INTERFACING CIRCUITRY CONSIDERATIONS

As with other display technologies, similar decode/driver and latching circuits used for other display technologies are used for interfacing between the output of a digital logic system and the liquid crystal display; many displays are available with onboard integrated circuitry. Because of the extremely low power that is needed, CMOS technology interface circuits are preferred. A direct drive approach is used for up to a 4-digit display with a multiplexing technique used for larger readouts. As the multiplexing rate increases, the operating temperature range becomes more critical and the viewing angle decreases.

# SPECIFICATIONS – LIQUID CRYSTAL DISPLAYS

## PRODUCT DESCRIPTION AND DEVICE FEATURES

The information on a data sheet includes the type of display, number of digits (or segments), color, font style, size and capabilities of the display. In addition, the characteristics and applications of the display are given, as well as any pertinent details for proper operation.

## PACKAGE OUTLINE

The drawing of the package appears on the data sheet and includes information on package dimensions, character height and width, mounting details, terminal spacing, terminal designations and lead material.

## ABSOLUTE MAXIMUM RATINGS
(At 25°C ambient temperature, unless otherwise noted)

### APPLIED VOLTAGE ($V_{ACmax}$)
The maximum A.C. voltage that can be applied between the contacts of the display pattern and the common back plate electrode. Typical values range from 12 to 35 volts.

### MAXIMUM CURRENT ($I_{max}$)
The total maximum current when all segments or dot-matrix cells are in the ON state. Typical values range from 10 to 100 microamperes.

### OPERATING TEMPERATURE RANGE
Generally, this range extends from 0°C to +70°C, however, LCD devices are available that range from −20°C to +80°C. Some direct drive devices range from −40°C to +85°C.

### STORAGE TEMPERATURE RANGE
Generally, from −40°C to +60°C, however, some devices can be stored in the range of −40°C to +85°C without damage.

### MAXIMUM OPERATING FREQUENCY
This value ranges from 30 to 120 Hertz.

### POWER DISSIPATION ($P_D$)
Depending on the device, values range from 100 microwatts to 10 milliwatts. With such low levels of power dissipation, no provisions are necessary for heat sinking or cooling.

ESTIMATED LIFE
Typical values: from 50,000 to 100,000 hours (6 to 12 years).

## ELECTRICAL AND OPTICAL CHARACTERISTICS
### (At 25°C ambient temperature)

OPERATING VOLTAGE
The range of operation is between 1.5 to 35 volts A.C., making this display compatible with the low voltage supplies used in modern equipment. If a battery is used as the supply, its D.C. voltage must be changed to A.C. with the use of a D.C. to A.C. inverter (chopper circuit).

CURRENT (All segments ON)
These values range from 0.1 to 10 microamperes.

CONTRAST RATIO
The measurement of the degree of difference between the light reflected by the energized segments of the display compared to the light reflected by, or transmitted through, the background field. A typical LCD contrast ratio is 20:1.

SPEED OF RESPONSE
Although LCD devices are relatively slow compared to the other display technologies, they are sufficiently fast to use multiplexing drive techniques. Typical values for turn-on and turn-off times range from 25 to 100 milliseconds. As ambient temperature decreases, response time increases, often to a value too high for multiplexing applications.

VIEWING ANGLE
This parameter is measured at a specified applied voltage while mounted in a vertical plane. Typical horizontal viewing angles range between 70° to 95°, with vertical viewing angles at about 60°. Several packaging techniques have been used to increase the horizontal viewing angle to about 120°, but generally the viewing angle for LCDs is much narrower than for all other display technologies.

Electrical characteristics of the latching and decode/driver circuits are generally specified in a data sheet. Typical and maximum values of supply current and logic input currents for its ON and OFF states are listed for the logic circuits at a specified supply voltage. In addition, any information required for blanking, timing or other specific operations is included.

# REINFORCEMENT EXERCISES

Answer TRUE or FALSE to each of the following statements:

1. Where good readability in high ambient light and/or viewing from long distances is required, the gas discharge display would be an appropriate selection.

2. The gas discharge display is sometimes referred to as a *planar* or *plasma* gas discharge or a *cold cathode* display.

3. Because a screen printing technique is used on gas discharge displays, it is difficult to display intricate patterns, such as bar graphs, both linear and circular, or special symbols or graphics.

4. A gas discharge display operates on the principle of emission of light from a gas caused by current flowing through the gas. To produce current flow, the ionization potential or breakdown voltage of the neon gas must be exceeded to ionize the gas. This reduces its resistance, thereby allowing current to flow.

5. Low D.C. voltage is required for energizing a gas discharge display, making it readily suitable for use in equipment using a low voltage (5 to 12 volts) power supply.

6. Gas discharge displays can only emit a bright orange glow.

7. For optimum operation of a gas discharge display, a suitable nonreflective matte filter is supplied with the display package to enhance contrast ratio.

8. Vacuum fluorescent technology displays are not easy to see under high ambient light conditions.

9. Vacuum fluorescent indicator panel displays require a separate low voltage for their filaments in addition to a high voltage supply for proper operation.

10. Vacuum flourescent displays can only emit a bluish-green light.

11. A liquid crystal display is generally black or gray. Other colors are available with the use of different filters.

12. Liquid-crystal displays (LCD) are extremely low-power, low-voltage readouts which operate on a principle of scattering light from an external source to provide a desired readout.

13. Since LCDs consume extremely low power, they are most appropriate for use in battery-operated equipment, including digital watches, hand-held calculators, portable computer monitors, digital thermometers and portable digital meters.

14. Liquid crystal displays will "wash-out" in direct sunlight and are packaged with appropriate shielding to protect the display from the direct rays of the sun.

15. LCD font style and size can be customized for any unique application since the display patterns are easily etched on the inner surface of one of its glass plates.

16. Liquid crystal displays can be designed for front lighted (reflective), back lighted (transmissive) or both modes of operation (transflexive).

17. The energizing voltage for a liquid crystal display can be either D.C. or A.C.

18. The viewing angle of an LCD is the widest of all display technologies.

19. Contrast ratio in a LCD is the measurement of the difference in the ambient light reflected (or scattered) by an energized display pattern, when compared to the light reflected by, or transmitted through, its background field. Contrast ratio is used to provide a quantitative measure of the quality of the display.

Answers to the reinforcement exercises are on page 327.

CHAPTER
THIRTEEN

# OPTO-COUPLERS (OPTO-ISOLATORS)

# OPTO-COUPLERS (OPTO-ISOLATORS)

An opto-coupler (opto-isolator) is an electronic component that transfers an electrical signal or voltage from one part of a circuit to another, or from one circuit to another, as it electrically isolates them from each other. It consists of an infrared emitting LED chip that is optically in-line with a light-sensitive silicon semiconductor chip, all enclosed in the same package. The silicon chip could be in the form of a photodiode, phototransistor or photo SCR.

## FUNCTIONS OF AN OPTO-COUPLER

- To isolate a voltage level in one section of a circuit from a different voltage level in another section.

- To prevent electrical noise or other voltage transients that may exist in one section of a circuit from interfering with a second section of the circuit when both sections are using a common circuit reference.

Noise or voltage transients can be caused by a poorly designed common reference, creating a "ground loop" condition. This problem can easily be eliminated with the use of an opto-coupler.

### SPECTRAL RESPONSE OF SILICON

Since silicon has a response to light (spectral response) that peaks at infrared wavelengths (between 800 and 950 nanometers), silicon devices are used in the opto-coupler as the photodetector section in conjunction with an infrared LED emitter. See Figure 131. Matching the infrared LED to the silicon chip achieves a maximum transfer of the desired electrical signal.

Different types of opto-couplers have specific characteristics that determine suitability for each unique application. The basic, and simplest type, is the opto-coupler with a photodiode used as its output section. The output of the opto-coupler is often connected to an amplifier to convert a low level voltage into a usable higher signal level.

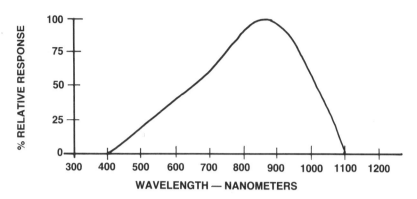

Spectral Response Curve of Silicon
**Figure 131**

# CONSTRUCTION AND PRINCIPLE
# OF OPERATION

The input section is an infrared LED chip separated from the silicon diode chip by a thin transparent mylar plate embedded in clear silicone (a derivative of silicon). The assembly is sealed in a package that is keyed to designate pin #1. The most commonly used opto-coupler package is the 6-pin plastic DIP (dual-in-line package). See Figures 132a and 132b.

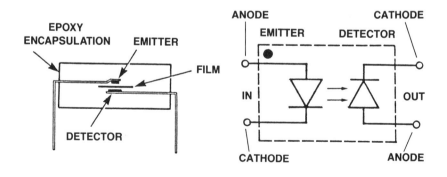

**Figure 132a**
Opto-coupler Cross-section

**Figure 132b**
Opto-coupler Schematic

When a forward bias voltage is applied to the input terminals of the LED (positive to the anode), an input current, $I_{IN}$, will flow in the LED circuit limited by the series resistor, $R_S$. The current produces infrared light at about 900 nanometers that is emitted from the LED and impinges on the photo-sensitive silicon chip.

## OPTO-COUPLER WITH PHOTODIODE OUTPUT SECTION

With light hitting the silicon diode in Figure 133, the photo-voltaic effect will cause light current to flow in the output section. With a load resistor, $R_L$, connected to the output terminals of the coupler, the light current, $I_{OUT}$, will develop a voltage, $V_L$, across the load. $V_L = I_{OUT} \times R_L$.

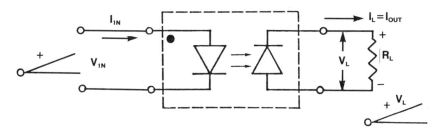

Opto-coupler with Photodiode Output Section
**Figure 133**

As the input signal varies, it causes the intensity of the infrared light to change. The output current, $I_L$, will also change, causing the output voltage, $V_L$, to change in the same manner. (As output current increases, output voltage will also increase, and vice-versa). A small change in input current will produce a proportionate change in output current. This characteristic of the opto-coupler makes it possible to couple either low-level analog signals or small D.C. voltage variations with little distortion.

In the circuit of Figure 133, both signal coupling and input-to-output isolation is achieved, however, the *current transfer ratio* (CTR) of a diode output opto-coupler is extremely low – about 0.10% to 0.15%.

The term *current transfer ratio* (CTR) defines the relationship of output current, $I_{OUT}$, to input current, $I_{IN}$.

$$CTR = \frac{I_{OUT}}{I_{IN}}$$

The output voltage, $V_L$, can be coupled to the input of an amplifier to increase its amplitude to a usable level.

The input section of all opto-couplers is an infrared LED, however, the output section can be different depending on the required application. The basic principle of operation is the same, regardless of the particular output section used.

## OPTO-COUPLER WITH PHOTOTRANSISTOR OUTPUT SECTION

Since the CTR of an opto-coupler with a photodiode output is so low, (0.10 to 0.15%), the silicon diode chip can be replaced with a silicon bipolar phototransistor. See Figure 134.

The transistor, with its inherent current gain, $\beta$, will provide a considerably higher CTR (between 20% and 100%) depending on the beta of the phototransistor.

The base lead of the transistor can be reverse biased to reduce sensitivity, or be forward biased to increase sensitivity or be left disconnected (floating), if desired.

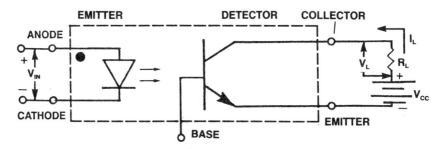

Opto-coupler with a Phototransistor Output Section
**Figure 134**

## OPTO-COUPLER WITH A PHOTO DARLINGTON OUTPUT SECTION

If a still higher CTR is desired, the bipolar transistor can be replaced with a Darlington transistor to serve as the photo-detector. See Figure 135.

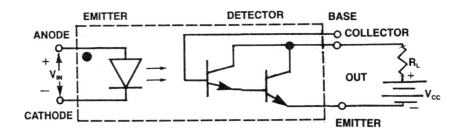

Opto-coupler with a Photo Darlington Output Section
**Figure 135**

In the circuits of Figures 134 and 135, the output current of the single bipolar phototransistor or the photo Darlington will develop an output voltage, $V_L$, across the load resistance, $R_L$. This voltage is the product of the output current, $I_C$, and the load resistance, $R_L$.

The opto-coupler can be operated either as a linear amplifier or as a digital switch, depending on the forward bias voltage applied to the base of the transistor.

OPTO-COUPLER WITH A PHOTO SCR OUTPUT SECTION
If the output section of the opto-coupler is a photo SCR, the coupler is normally used for switching A.C. in its output, operating under the same principle as an ordinary SCR circuit.

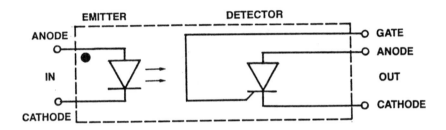

Opto-coupler with a Photo SCR Output Section
**Figure 136**

The SCR gate current is achieved through the photo-voltaic action produced by infrared light impinging on the SCR gate, while isolating the two circuits at the input and output of the opto-coupler.

Steady-state D.C. can be used as the supply voltage, causing the output to be latched ON when the SCR is energized. This type of circuit is typically used for security or fire alarm systems. To turn the system OFF after being triggered, a SPST, normally-closed switch for circuit reset can be inserted in the supply voltage path to momentarily remove the supply voltage after the input pulse returns to zero. This action allows the SCR to turn OFF and return to its standby state.

The same techniques for SCR turn-off, discussed in CHAPTER FIVE — THYRISTORS, can be used with photo SCR opto-couplers.

## SLOTTED OPTO-COUPLER

The slotted opto-coupler is available with both phototransistor and photo Darlington photodetectors, with the device package structured to provide an additional element of control. The package has an air gap approximately one-sixteenth of an inch in length between its two sections. One section has an infrared LED and the other, a photodetector. See Figure 137.

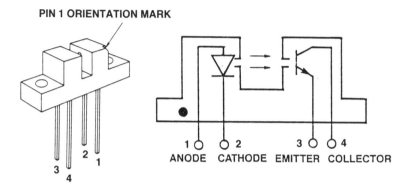

Slotted Opto-coupler Package and Schematic
**Figure 137**

# SLOTTED OPTO-COUPLER APPLICATIONS

The infrared light beam that links the two sections of an opto-coupler can be broken by the mechanical insertion of any thin object into its air gap that can block infrared light. This device lends itself to many control applications.

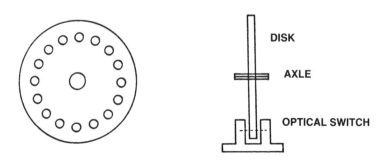

Slotted Opto-coupler Application
**Figure 138**

When a thin plastic or metal disk, with perforations or slots cut near or at its outside edge (see Figure 138), is rotated inside the slot of the opto-coupler, the infrared light can be detected when there is an opening at the slot. As a result, current flows at the output of the coupler. When the infrared light is blocked by the opaque section of the disk, there is no output current.

As the infrared light beam is interrupted, pulses are generated at the output of the coupler and rotational speed of the disk can be measured. If the disk is rotated by the flow of a liquid, the rate of flow of the liquid can be determined. For example, with a properly calibrated counting system, the exact number of gallons of gasoline being pumped at a gas station can be measured accurately with the use of this device.

Other applications of a slotted opto-coupler include a punched-card reader, parts counter, end of tape sensor on a printer or tape recorder and as part of a safety interlock mechanism on an equipment cabinet door.

In addition to the types already discussed, opto-couplers are manufactured with many variations of their photodetector sections. One type consists of an infrared LED emitting light to an integrated photodiode/bipolar transistor circuit, allowing direct coupling to digital logic circuitry. See Figure 139a. Another type uses a photodiode directly coupled to a logic circuit as the output section. See Figure 139b.

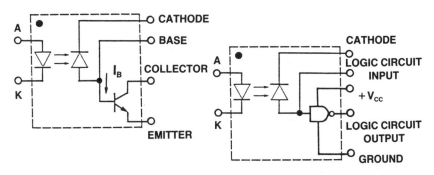

**Figure 139a**                    **Figure 139b**
Variations of Opto-coupler Output Sections

Opto-couplers are available not only as single components, but as duals (two separate devices in one package) and also as quads (four separate devices in one package).

# OPTO-COUPLER SPECIFICATIONS

## PRODUCT DESCRIPTION AND DEVICE FEATURES

Information on coupler type, circuit configuration, linearity, frequency response, switching speed, isolation voltage and any typical applications is listed on the data sheet.

## PACKAGE OUTLINE

The drawing of the package appears on the data sheet and includes information on package dimensions, mounting details, terminal spacing, terminal designations and lead material.

## ABSOLUTE MAXIMUM RATINGS
### (At 25°C ambient temperature)

ISOLATION VOLTAGE

The maximum voltage differential the device can tolerate between its input and output sections. Isolation voltage depends on the insulation used – air, glass or plastic. Typical values of isolation voltage range between 500 and 6000 volts.

OPERATING AND STORAGE TEMPERATURE

Plastic packages range from $-55°C$ to $+100°C$. Hermetically sealed devices range between $-55°C$ to $+125°C$.

LEAD SOLDERING TEMPERATURE

Typical value is 230° for 7 seconds with the soldering spot at least ¼ inch from the seal between lead and package.

## INPUT SECTION (INFRARED LED)

These specifications are identical to those which are defined and discussed in CHAPTER TEN – LIGHT EMITTING DIODES and listed under the heading of ABSOLUTE MAXIMUM RATINGS. They include:

- PEAK INVERSE (REVERSE) VOLTAGE (PIV) or (PRV)
- REVERSE D.C. CURRENT ($I_R$)
- CONTINUOUS D.C. FORWARD CURRENT ($I_{Fcont}$)
- PEAK FORWARD CURRENT ($I_{Fpeak}$)
- D.C. POWER DISSIPATION ($P_D$)

## DETECTOR SECTIONS

These specifications are identical to those which are defined and discussed in previous chapters and listed under the heading of ABSOLUTE MAXIMUM RATINGS as follows:

PHOTODIODE
CHAPTER THREE – DIODE SPECIFICATIONS. Ratings include:
- REVERSE VOLTAGE ($V_R$)
- D.C. POWER DISSIPATION ($P_D$)

PHOTOTRANSISTOR AND PHOTO DARLINGTON
CHAPTER SIX – BIPOLAR TRANSISTORS. Ratings include:
- COLLECTOR-TO-EMITTER VOLTAGE ($V_{CEmax}$)
- COLLECTOR-TO-BASE VOLTAGE ($V_{CBmax}$)
- EMITTER-TO-BASE VOLTAGE ($V_{EBmax}$)
- D.C. POWER DISSIPATION ($P_D$)

PHOTO SCR
CHAPTER FIVE – THYRISTORS. Ratings include:
- REPETITIVE PEAK REVERSE (INVERSE) VOLTAGE (PRV) or (PIV)
- NON-REPETITIVE REVERSE VOLTAGE ($PRV_{TRANSIENT}$)
- PEAK POSITIVE ANODE VOLTAGE (PFV)
- PEAK ONE-CYCLE SURGE CURRENT ($I_{SURGE}$)
- AVERAGE FORWARD CURRENT ($I_F$)
- PEAK ONE-CYCLE SURGE CURRENT ($I_{SURGE}$)
- PEAK GATE VOLTAGE ($V_{GM}$)
- PEAK POSITIVE GATE CURRENT ($I_{GM}$)
- TOTAL POWER CAPABILITY ($P_T$)

## ELECTRICAL CHARACTERISTICS
## INPUT SECTION (INFRARED LED)

These specifications are identical to those which are defined and discussed in CHAPTER TEN – LIGHT EMITTING DIODES and listed under the heading of ELECTRICAL CHARACTERISTICS. They include:
- FORWARD VOLTAGE ($V_F$)
- REVERSE LEAKAGE CURRENT ($I_R$)

## DETECTOR SECTION

These specifications are identical to those which are defined and discussed in previous chapters and listed under the heading of ELECTRICAL CHARACTERISTICS as follows:

PHOTODIODE

CHAPTER THREE-DIODE SPECIFICATIONS. Characteristics include:

- FORWARD VOLTAGE ($V_F$)
- REVERSE OR LEAKAGE CURRENT ($I_L$)
- TURN-ON TIME ($T_{ON}$) AND TURN-OFF TIME ($T_{OFF}$)

PHOTOTRANSISTOR AND PHOTO DARLINGTON

CHAPTER SIX – BIPOLAR TRANSISTORS. Characteristics include:

- COLLECTOR CURRENT CUTOFF (LEAKAGE) ($I_{CES}$)
- CURRENT GAIN – BETA ($\beta$) or ($h_{FE}$)
- COLLECTOR SATURATION RESISTANCE ($R_{CE(sat)}$)
- TURN-ON TIME ($T_{ON}$) AND TURN-OFF TIME ($T_{OFF}$)
- FREQUENCY CUTOFF ($F_{CO}$)

PHOTO SCR

CHAPTER FIVE – THYRISTERS. Characteristics include:

- GATE TURN-ON VOLTAGE ($V_G$) AND GATE TURN-ON CURRENT ($I_{GT}$)
- FORWARD VOLTAGE ($V_F$) AND REVERSE CURRENT ($I_R$)
- TURN-ON AND TURN-OFF TIMES ($T_{ON}$) and ($T_{OFF}$)

## TRANSFER CHARACTERISTICS

CURRENT TRANSFER RATIO – CTR

The term *current transfer ratio* (CTR) defines the relationship of output current, $I_{OUT}$, to input current, $I_{IN}$, and is measured as a percentage of the input current. Typical values are:

- Photodiode output – from 0.10% to 0.15%
- Phototransistor output – from 20% to 100%
- Photo Darlington output – from 100% to 1000%

The output of an opto-coupler can be connected to any other circuit with the entire assembly enclosed in a single package and treated as a single component. This approach lends itself to the creation of many products that include opto-coupling as part of their characteristics. Regardless of the circuit design, the same operating principle is maintained – optical-coupling between circuits while achieving electrical isolation between these circuits. See Figure 140.

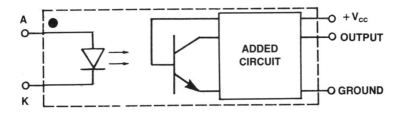

Opto-coupler with Added Circuit
**Figure 140**

## REINFORCEMENT EXERCISES

Answer TRUE or FALSE to each of the following statements:

1. An opto-coupler, or opto-isolator is used to isolate a circuit connected to its input section from a different circuit connected to its output section.

2. Electrical noise, or unwanted voltage, that exists between parts of a circuit reference can not be easily eliminated with the proper use of an opto-coupler.

3. An opto-coupler consists of an infrared LED input section and an infrared-sensitive photodetector output section.

4. Any photodetecting material is applicable for use as the output section of an opto-coupler.

5. Opto-couplers are available in a variety of output section choices. The different output sections include: photodiode, phototransistor, photo Darlington and photo SCR.

6. The CTR, or current transfer ratio, (output current to input current), of an opto-coupler using a photodiode as the output section is close to 100%.

7. The CTR of an opto-coupler using a photo Darlington output section is 100% or greater.

8. The non-linear characteristic of an opto-coupler with a photodiode output section makes it impossible to couple low-level analog signals or small D.C. voltage variations with little distortion even with a small change in input current.

9. To achieve very high current transfer ratio values, an opto-coupler with a photo Darlington output section is used.

10. If the output section of the opto-coupler is a photo SCR, the coupler is normally used to switch A.C. in its output circuit, with the photo SCR operating by the same principle as an ordinary SCR.

11. Steady-state D.C. is never used as the supply voltage for an opto-coupler with a photo SCR output section, since the photo SCR will be latched in its ON state after being energized, even if the photo SCR gate current is removed.

12. The slotted opto-coupler provides an additional element of control by using a package that has a slot between its two sections. In blocking the infrared light that optically links the sections, the transfer of the signal coming from the input is interrupted.

13. The maximum voltage differential allowed across the gap between input and output sections of an opto-coupler is called isolation voltage and is limited to a maximum of 100 volts.

14. Opto-couplers are available as single components, as two separate devices in one package (dual) and as four separate devices in one package (quad).

Answers to the reinforcement exercises are on page 329.

CHAPTER
FOURTEEN

# SOLID-STATE RELAYS (SSR)

### PRINCIPLE OF OPERATION

### COMPARISON WITH ELECTROMECHANICAL RELAYS

### SSR SPECIFICATIONS

### REINFORCEMENT EXERCISES

# SOLID STATE RELAYS (SSR)

A solid-state relay (SSR) is the semiconductor version of an electromechanical relay and is controlled by an infrared optocoupler to produce isolation between its actuating and switching sections. When used with additional circuitry, it provides many of the features of relay switching without the problems that are inherent in electromechanical relays. The SSR is manufactured as a hybrid circuit and is treated as a single component.

## HOW THE SSR FUNCTIONS

In the electromechanical relay shown in Figure 141a, an actuating input current will cause the contacts of the switching section to close. Removing the actuating current will allow the switch to return to its non-energized state. In comparison, the SSR functions as follows:

- The infrared LED in the circuit of Figure 141b is energized by applying a relatively small D.C. voltage to the input terminals. If A.C. voltage is used to energize the LED, it must be rectified and filtered with an appropriate circuit.

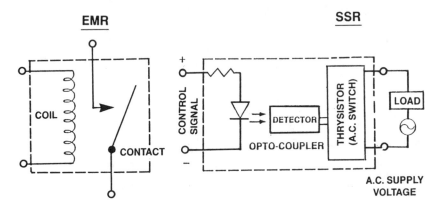

**Figure 141a**               **Figure 141b**

Electromechanic Relay Compared with the SSR

- The infrared light is opto-coupled to an appropriate photo-detector (either a phototransistor or a photo Darlington), whose output current provides the input to a power thyristor (SCR or TRIAC) gate terminal.

- With gate current flowing in the thyristor, load circuit will flow when the A.C. supply voltage is present – positive for an SCR and positive or negative for a TRIAC.

  As long as the input voltage is maintained to keep the infrared LED energized, the SSR will remain ON. Either pulsating D.C. voltage will be across the load of an SCR or A.C. voltage will be across the load of a TRIAC.

- When the input control signal is removed, the infrared LED output reduces to zero, reducing the thyristor gate current to zero.

- When the A.C. supply voltage reaches its zero value, the thyristor will turn OFF and no current will flow in the load. As long as the input (control) current remains at zero, the SSR will remain OFF.

Even though the SSR in the circuit of Figure 141b was turned OFF at the zero-crossing point of the A.C. supply voltage, the thyristor could have turned ON at any level of the A.C. supply voltage.

To prevent the sudden application of output current that might produce an unwanted voltage transient or radio frequency interference (RFI), it is desirable to turn the SSR ON, as well as OFF, at the zero-crossing point of the A.C. supply voltage.

This can be accomplished by adding a sensing/trigger circuit, called a "Schmitt trigger", to the original SSR circuit shown in block form in Figure 141b.


SCHMITT TRIGGER
The Schmitt trigger is energized (enabled) by applying a current to its input. Once enabled, it will sense the condition of the A.C. supply voltage and react to the level of the supply voltage in the following way:

- As long as the A.C. voltage remains at an instantaneous voltage other than zero, the output current of the Schmitt trigger will be zero.

- When the A.C. voltage passes through the zero voltage level, a current will exist at the output of the Schmitt trigger and remain there until the trigger is disabled.

  To disable the Schmitt trigger, the current at its input must be reduced to zero. Once disabled, the Schmitt trigger will not respond to any level of A.C. supply voltage and its output will be at zero.

The sequence of operation of the SSR can now be observed with the addition as shown in Figure 142. Since current is moving at the speed of light, the individual functions occurring at the specific time points are happening instantaneously.

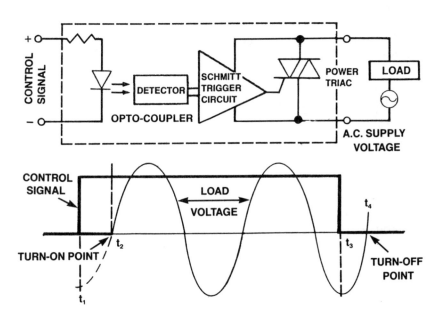

Opto-isolated Zero-crossing SSR
**Figure 142**

- At time $T_1$, a positive D.C. step voltage is applied to the input terminals of the SSR, energizing the infrared LED.

- The infrared light, impinging on the photodetector, causes current to flow in its input section (photo-voltaic effect), which produces a photodetector output current. This output current is also the input current to the Schmitt trigger.

- The input current enables (energizes) the Schmitt trigger, allowing it to sense the condition of the A.C. supply voltage. At this point in time, $T_1$, the A.C. supply voltage is at some value other than zero and the output of the Schmitt trigger is zero, therefore, the gate current of the TRIAC is also zero and the TRIAC remains OFF.

- As the A.C. supply voltage continues to move, it reaches the zero voltage point at time $T_2$. It is at this point that the Schmitt trigger, sensing zero A.C. supply voltage, produces current in the gate circuit of the TRIAC.

- With TRIAC gate current present, as the A.C. supply voltage leaves zero, load current will flow, turning the SSR ON.

- At time $T_3$, the D.C. input voltage is reduced to zero, de-energizing the LED, reducing the output of the photodetector to zero, that, in turn, causes the Schmitt to be disabled. Disabling the Schmitt trigger reduces the TRIAC gate current to zero, however, this does not turn the TRIAC OFF.

- At time $T_4$, the A.C. supply voltage crosses the zero level point and the TRIAC turns OFF. Since there is no current flowing in its gate circuit, the TRIAC stays OFF and, at this point, the SSR is turned OFF.

Note that the SSR was turned ON and turned OFF at a point of zero A.C. supply voltage, eliminating the generation of either RFI or voltage transients normally created in switching current in an inductive load. In addition to its opto-isolation capability, zero-crossing is the added feature that provides the full name to the SSR – an opto-isolated, zero-crossing, solid-state relay, generally, just called SSR.

Only A.C. voltages can be switched with this type of SSR (using either an SCR or TRIAC as the switching device). Other types of SSR output devices can be used – bipolar transistors and power MOSFFTs. Depending on the output circuit configuration, they can be used as follows:

- Bipolar transistor – for switching D.C. at high current or both A.C.and D.C. at low levels (less than 100 milliamperes).

- Power MOSFET – a single device is used for D.C. switching at high voltage and current levels – 400 volts at 40 amperes. By combining two power MOSFETs, an SSR may be used to switch both A.C. and D.C. for analog switching applications.

# COMPARISONS WITH ELECTROMECHANICAL RELAYS

ADVANTAGES:
- There are no moving parts, therefore, no mechanical wear or friction.

- An SSR has no contact bounce, eliminating the possibility of circuit instability.

- Operation from logic level voltages to provide a direct interface between a computer output and an A.C. switching load. The actuating voltage and current requirements are very low – from 3 to 35 volts D.C. at 1 to 10 milliamperes.

- SSRs are inherently arc-free, since they have no physical contacts across which an arc can form.

- Since there is no arcing of contacts, there is no generation of radio frequency interference (RFI).

- There is no "relay clack" or audible contact noise.

- SSRs provide zero-crossing ON and OFF switching operation to suppress voltage transients that would otherwise be produced by switching inductive loads. A *snubber network* (transient suppressor network), consisting of a resistor and capacitor in series, is generally incorporated across the output section of the SSR to minimize the effects of load-generated stresses on the output section. See Figure 143.

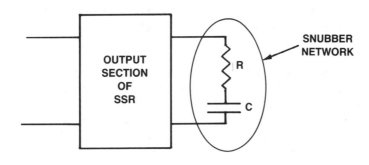

Typical Snubber Network
**Figure 143**

- Even with the inherent zero-crossing time delay, switching action is faster than slower, "run-of-mill" electromechanical relays that usually take hundreds of milliseconds to switch.

- An SSR provides large switching power capability in a relatively small package – high power SSRs are capable of switching 40 amperes at 400 volts A.C. in a package about half the size of a comparable electromechanical relay.

- An SSR can be mounted in any plane or position.

- An SSR requires no maintenance, periodic or otherwise, since there are no contacts to burnish or replace.

- Solid-state relays are resistant to shock, vibration and the corrosive environments of chemical solvents and fungus, with no need for special precautions.

- The device has the inherent reliability of semiconductors, offering infinite life expectancy.

DISADVANTAGES:
- The original SSRs did not have multiple-pole, multiple-throw contact capability, normally available in electromechanical relay contact configurations. Newer SSR designs, however, are incorporating more contact configuration capability, including double-pole, double-throw (DPDT) contact forms.

- SSRs have an inherently higher resistance across their "contacts" than an electromechanical relay. Because of this condition, the SSR generates more heat ($I^2R$), particularly at higher currents. Since the SSR dissipates higher power, it is not as temperature-tolerant as an electromechanical relay and some provision must be made for cooling. Adding cooling components adds more cost to the overall system.

- In some applications, totally disconnecting a load, normally provided by an electromechanical relay, may be necessary. For some SSRs, having an output leakage current as high as several milliamperes, it may be required to add other components, such as a switch or circuit breaker in series with the SSR output, to assure a total disconnect of the load.

- Initial cost is higher for an SSR than for an electromechanical relay with comparable specifications.

# SSR SPECIFICATIONS

Since an SSR is basically a semiconductor circuit, its data sheet more closely resembles semiconductor specifications than electromechanical relay specifications.

## PRODUCT DESCRIPTION AND DEVICE FEATURES

The data sheet includes general information on the input and output voltage ranges, current capability, A.C. or D.C. operation (or both), and package type. In addition, it lists UL (Underwriters Laboratory), CSA (Canadian Standards Association) and VDE (European standards) approval or certification information, isolation voltage and any other features.

## PACKAGE OUTLINE

The drawing of the package appears on the data sheet and includes information on package dimensions, mounting details, terminal options, terminal spacing, terminal designations and lead material.

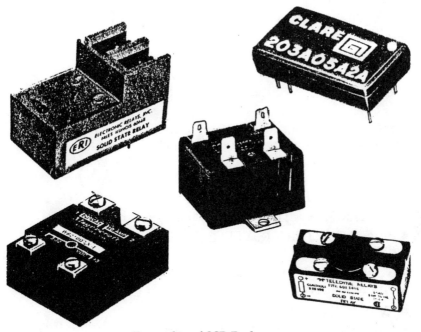

Examples of SSR Packages
**Figure 144**

## ABSOLUTE MAXIMUM RATINGS
(At 25°C case temperature, unless otherwise noted)

ISOLATION VOLTAGE
The maximum allowable voltage from input-to-output, input-to-case and output-to-case. These values range from 500 to 4000 volts A.C. at 60 Hertz, depending on the package.

OPERATING AND STORAGE TEMPERATURE
Military types – hermetically-sealed packages: From $-55°C$ to $+125°C$.
Commercial, consumer, industrial types – plastic packages: From $-40°C$ to $+80°C$.

## INPUT SECTION (INFRARED LED)

These specifications are identical to those which are defined and discussed in CHAPTER TEN – LIGHT EMITTING DIODES and listed under the heading of ABSOLUTE MAXIMUM RATINGS. They include:
- PEAK INVERSE (REVERSE) VOLTAGE (PIV) or (PRV)
- REVERSE D.C. CURRENT ($I_R$)
- CONTINUOUS D.C. FORWARD CURRENT ($I_{FDNT}$)
- PEAK FORWARD CURRENT ($I_{Fpeak}$)
- D.C. POWER DISSIPATION ($P_D$)

## OUTPUT SECTION

These specifications are identical to those which are defined and discussed in previous chapters and listed under the heading of ABSOLUTE MAXIMUM RATINGS as follows:

TRIAC OUTPUT
CHAPTER FIVE – THYRISTERS. Ratings include:
- REPETITIVE PEAK REVERSE (INVERSE) VOLTAGE (PRV or PIV)
- NON-REPETITIVE REVERSE VOLTAGE ($PRV_{TRANSIENT}$)
- PEAK POSITIVE ANODE VOLTAGE (PFV)
- AVERAGE FORWARD CURRENT ($I_F$)
- PEAK ONE – CYCLE SURGE CURRENT ($I_{SURGE}$)
- TOTAL POWER CAPABILITY ($P_T$)

BIPOLAR TRANSISTOR OUTPUT
CHAPTER SIX – BIPOLAR TRANSISTORS. Ratings include:
- COLLECTOR-TO-EMITTER VOLTAGE ($V_{CEmax}$)
- D.C. POWER DISSIPATION ($P_D$)

MOSFET OUTPUT
CHAPTER SEVEN – FETs. Ratings include:
- DRAIN-TO-SOURCE VOLTAGE ($V_{DS}$)
- DRAIN CURRENT ($I_{D(max)}$)
- POWER DISSIPATION ($P_D$)

# ENVIRONMENTAL SPECIFICATIONS

Shock, vibration and acceleration specifications are listed for military type solid-state relays. These values are generally defined by the appropriate military specifications.

### ELECTRICAL CHARACTERISTICS
(At 25°C case temperature unless otherwise specified)

### INPUT (CONTROL SECTION)

CONTROL VOLTAGE RANGE ($V_{IN}$) SSRs are specified for either D.C. or A.C. operation.

The range is specified as a minimum and maximum value and is typically from 3 to 35 volts D.C. for D.C. operated devices. This control voltage range makes the SSR compatible with logic voltage levels for direct coupling from standard digital logic circuits and is the most common mode of operation for the SSR.

For A.C. operated devices, the range is specified as a minimum and maximum value and is typically from 90 to 250 volts A.C. at a specified frequency (generally from 47 to 70 Hertz).

INPUT CURRENT ($I_{IN}$)
Control current requirements are generally low enough to be supplied from logic circuitry outputs. Typically, the input current ranges from 1 to 30 milliamperes.

MUST TURN-ON VOLTAGE
The minimum required input voltage, $V_{IN}$, at specified temperature limits to turn the device ON.

MUST TURN-OFF VOLTAGE The maximum input voltage at which point the device will turn OFF. Typical values at specified temperature limits:
For D.C. operation $-1$ volt.
For A.C. operation $-20$ volts.


## OUTPUT SECTIONS

These specifications are identical to those which are defined and discussed in previous chapters and listed under the heading of ELECTRICAL CHARACTERISTICS as follows:

TRIAC OUTPUT
CHAPTER FIVE – THYRISTERS. Characteristics include:
- REVERSE CURRENT ($I_R$)
- TURN-ON TIME ($T_{ON}$)
- TURN-OFF TIME ($T_{OFF}$)

BIPOLAR TRANSISTOR OUTPUT
CHAPTER SIX – BIPOLAR TRANSISTORS. Characteristics include:
- COLLECTOR CURRENT CUTOFF (LEAKAGE) $I_{CES}$)
- FREQUENCY CUTOFF ($F_{CO}$)
- COLLECTOR SATURATION RESISTANCE ($R_{CEsat}$)
- TURN-ON TIME – $T_{ON}$
- TURN-OFF TIME – $T_{OFF}$

MOSFET OUTPUT
CHAPTER SEVEN – FETs. Characteristics include:
- DRAIN-TO-SOURCE ON RESISTANCE ($R_{ON}$)
- DRAIN CUTOFF CURRENT ($I_{D(OFF)}$)
- DRAIN CURRENT IN ENHANCED STATE ($I_{D(ON)}$)
- DRAIN-TO-SOURCE CAPACITANCE ($C_D$)
- TURN-ON TIME ($T_{ON}$)
- TURN-OFF TIME ($T_{OFF}$)

## REINFORCEMENT EXERCISES

Answer TRUE or FALSE to each of the following statements:

1. A solid-state relay consists basically of three parts: an infrared LED opto-coupler input circuit, an output switching circuit and other associated circuitry to enhance its operation as a relay.

2. Solid-state relays only switch A.C. voltages at its output and only respond to D.C. control voltages at its input.

3. A Schmitt trigger is used in a solid-state relay to provide the zero-crossing turn-on feature.

4. In the operation of the Schmitt trigger, when an A.C. voltage is sensed at a value other than zero, there will be output current from the trigger. If zero A.C. supply voltage is sensed, the output of the Schmitt trigger will be zero.

5. In a TRIAC output SSR (used for A.C. voltage switching), removal of the gate current to the TRIAC will turn the SSR OFF regardless of the condition of the A.C. supply voltage.

6. With the zero-crossing characteristic of the SSR, no RFI signals or inductive load transients are generated.

7. Solid-state relays are available with a bipolar transistor in the output circuit for switching D.C. at high current levels or for switching both A.C. and D.C. at low output current levels.

8. A single power MOSFET can be used as the output of an SSR for switching D.C. at high voltage and high current levels.

9. By combining two power MOSFETs in the output, an SSR may be used to switch both A.C. and D.C. for analog switching.

10. The most important advantage of solid-state relays over electromechanical relays is that they are much quieter in their operation.

11. An advantage of the SSR is its ability to provide multiple switching contact configurations, whereas, only a single-pole, single-throw capability is provided by an electromechanical relay.

12. Because of its zero-crossing feature, the action of the SSR responds much slower than the ordinary electromechanical relay.

13. There is less heat generated in a solid-state relay than in an electromechanical relay having comparable performance specifications.

14. The initial cost of an SSR is higher than an electromechanical relay, however, there is no maintenance required for the SSR and the elimination of those maintenance costs must be considered in evaluating overall system long-term costs.

Answers to the reinforcement exercises are on page 331.

CHAPTER
FIFTEEN

# OPTOELECTRONIC TECHNOLOGY TRENDS

LARGE FLAT-PANEL DISPLAYS

- LIQUID-CRYSTAL DISPLAY TECHNOLOGY
- ELECTROLUMINESCENT DISPLAY TECHNOLOGY

FIBER OPTICS

# TECHNOLOGY TRENDS

## LARGE FLAT-PANEL DISPLAYS

Because of the increasing demand for portable computers, terminals and instruments, there has been a concomitant demand for enlarged and improved displays to satisfy readout requirements. Manufacturers of large flat-panel displays are meeting this market demand with refinements in technology and manufacturing procedures and are changing the nature of large displays. These innovative products are challenging the dominance of the long established cathode ray tube (CRT) for use as a video monitor.

The scope of large flat-panel displays encompasses two major areas of technology – liquid-crystal displays (LCD) and electroluminescent displays (EL), a variation of vacuum fluorescent display technology.

### LIQUID CRYSTAL DISPLAY (LCD) TECHNOLOGY

The use of the *supertwisted* liquid-crystal, commonly referred to as the SBE (supertwisted birefringence effect) LCD is a significantly advanced approach to display technology. SBE technology offers a distinct improvement in quality over the older nematic field-effect LCD by enhancing the display to allow the eye's ability to:

- Distinguish the uniformity of a display more clearly
- Focus more sharply on patterns shown on a uniform background
- Use its inherent sensitivity to certain colors

In an SBE LCD, the angle of rotation within the display cell is increased from the 90° used in the nematic field-effect LCD to 180° or more. The amount of twisting in the liquid-crystal and the orientation of the polarizer used in the supertwisted LCD creates a uniform yellow-green background. *Birefringence* is the splitting of light into two vertical waves to optimize the contrast in colors that the human eye perceives as dark and light. It improves the clarity of the display and creates a wider viewing angle than is possible with a nematic field-effect LCD. Increased twisting of the SBE liquid-crystal enhances the ability of the human eye to further distinguish the background uniformity of the display.

Because the colors of the display and background areas of the SBE LCD depend on the mechanical orientation of the polarizer axis in relation to the twist of the liquid-crystal, two types of super-twisted displays are possible. In one type, the display is dark blue with a yellow background and in the other, the display is grayish-white with a dark blue background. The second format, when enhanced by backlighting, is suitable for use as a reverse video display. With the use of both formats, the supertwisted liquid-crystal flat-panel display can be a worthy competitor to the mono-chrome CRT monitor.

The SBE LCD has established new standards for large flat-panel displays, however, the next step in the evolution of display tech-nology is the perfection of an active-matrix full color flat-panel dis-play. This could become the first display to emulate a full color CRT monitor.

Although many technological problems still exist, the approach to a full color display has been to deposit each individually address-able picture element, called *pixel*, on the panel. Every pixel area has a component created on a glass substrate by using a microthin semiconductor process. In addition, every pixel has an associated thin-film transistor (TFT) on the same glass substrate to be used as the driver. When TFT technology eventually becomes practical, complete logic systems, including microprocessors and associated circuitry, will be incorporated in patterns on the glass panel sur-face. This capability will eliminate the need for PC board inter-connects, offering more efficient use of space and greater circuit reliability.

The process used to make a TFT LCD mask is basically the same as that used for an semiconductor integrated circuit. This process involves the diffusion of several patterns with the addition of a three-step color-filter screen on top of each pixel. If a transistor fails, the pixel will not light, producing a black spot on the panel. If an internal interconnect burns out, all the pixels in its activating area will remain dark, resulting in a black line on the panel. Redundant patterns are created in an attempt to minimize the effects of these failures. At the present time, electrical stability of thin-film transistors can also be upset by the stresses of shock, vibration, light and temperature.

In the very near future, improvements in TFT LCD technology will produce large flat-panel displays that will provide a viable alternative to color CRT monitors for many applications.

## ELECTROLUMINESCENT (EL) DISPLAY TECHNOLOGY

For many years, cathode ray tubes (CRTs) have set the standard for computer monitors, radar displays and similar equipment. They provide bright, full color, they can be dimmed uniformly, show gray shades, have good contrast and can be viewed at very wide angles. They are also very bulky, heavy, use very high voltages (greater than 10 KV for full color CRTs) and consume a great deal of power. In addition, since CRTs are really vacuum tubes, they are not shock and vibration proof and suffer from problems common to all vacuum tubes – they emit large amounts of unwanted heat and have a relatively short life expectancy.

With the advent of electroluminescent (EL) displays, most of the benefits of the CRT can be achieved without inherent disadvantages. EL technology produces an aesthetically pleasing amber display that can be modified to provide a full color capability. It is the only display technology, besides the CRT, that provides brightness and contrast controls and, in addition, it has the features of semiconductor reliability. EL displays are being used in applications other than replacements for portable computer monitors. Since they need to be dimmed at night, they are suitable for use as the display panels of automobiles, airplanes and ships.

Thick-film electroluminescent technology is primarily aimed at lamp and illumination applications (backlighting for instrument panels) rather than for information displays. This approach uses a pulverized phosphor solid that is pressed into a plastic or ceramic binder. When an A.C. voltage is applied across the electrodes of the device, it will emit a bright amber glow.

Thin-film electroluminescent (TFEL) technology uses electrobeam sputtering, vacuum deposition or thermal evaporation of phosphor material to provide a highly uniform phosphor layer. The TFEL approach produces large-panel displays capable of favorably competing with other technologies in data-display and graphics-display applications.

A TFEL display is built around a thin-film electroluminescent phosphor that is sandwiched between two electrodes placed at right angles to each other. The combination of zinc sulphide phosphor doped with manganese (ZnS:Mn) produces amber (orange-yellow) light. Theoretically, altering the amount of doping material can produce other colors – green, blue, red and white, which will eventually provide full color spectrum displays. Athough all manufacturers use the same basic structure and the same basic phosphor, the electrode and insulating-layer materials may vary. Two typical cross-section views of a thin-film electroluminescent display structure are shown in Figure 145.

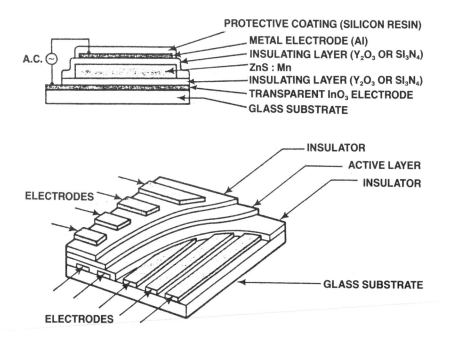

Two Cross-section Views of a TFEL Display Structure
**Figure 145**

- The phosphor (the light emitting layer) is sandwiched between two insulating layers made of yttrium oxide ($Y_2O_3$) or silicon nitride ($Si_3N_4$).

- An aluminum (Al) electrode is on top of one insulation layer. A transparent layer of conducting indium oxide is placed beneath the other insulating layer and forms a monolithic capacitor structure.

- This structure is then placed on a glass substrate and its external contacts are connected to the two electrodes.

- The entire structure is covered with a silicone resin sealant to prevent contamination and protect it against humidity.

When an A.C. voltage above the threshold value, $V_T$, (typically 150 to 200 volts at a frequency between 2 to 5 kilohertz) is applied across the selected pair of electrodes, the phosphor at that pixel intersection is energized, producing an amber colored light dot.

The display has an extremely high contrast turn-on because the device emits less than 1% of its ON brightness at points where the applied A.C. voltage is below the threshold voltage level. Brightness increases very sharply beyond the threshold voltage. See Figure 147. The applied A.C. voltage should be limited to a value below the saturation voltage, $V_S$, (typically 300 volts) to prevent permanent damage to the display.

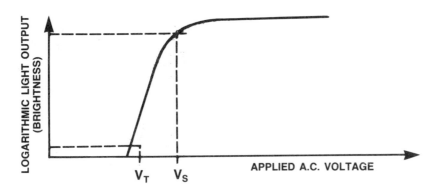

Light Output of a TFEL Display
**Figure 146**

No light is emitted outside of the region directly between the energized electrodes because the voltage at the non-emitting areas is below the threshold value, $V_T$, of the applied A.C. voltage. These non-emitting sections experience only fractions (about 1%) of that voltage.

Thin-film electroluminescent displays can be produced with very large-area configurations and are typically twice as bright as a standard black and white television set display. The relationship between brightness and voltage is shown in Figure 146.

As the amplitude of the A.C. voltage increases, light output increases logarithmically. Increasing the voltage above the saturation voltage, $V_S$, eventually causes voltage breakdown and panel damage. TFEL displays maintain constant brightness across the panel, dim uniformly and are viewable in direct sunlight. Room-light contrast ratios of 30:1 to 50:1 are typical.

Despite their advantages, TFEL displays do have some problems. The current levels required to charge and discharge the inherent capacitor cells every cycle can be 10 times as much as the dissipative currents required for the production of light. For this reason, a large, costly power supply is required.

In addition to power supply costs, the overall manufacturing expense of EL panels have traditionally been a drawback, since each display line needs its own driver. With high volume production, the basic panel costs can be reduced since the manufacturing process is fairly simple – TFEL panels require only two lithographic steps. As TFEL panels become more commercially acceptable, prices will drop.

Generally, it is felt that the key to practical TFEL displays is the availability of high voltage drivers on a single circuit board or module. These boards or modules will be designed to take standard logic display signals and deliver the properly timed high voltage pulses needed by the panel. In this manner, the TFEL display can be integrated into an existing system with no need for major modifications to the system.

To achieve sufficient brightness, a relatively large voltage (150 to 200 volts A.C. at a frequency between 2 to 5 kilohertz) is required. This range of voltage is not compatible with existing monolithic silicon technology, particularly for large-matrix driver configurations. The matrix format is ideal for many types of displays, including alphanumeric text, graphics and video with gray scale. Attempts have been made to lower the operating voltage of the TFEL display, however, reduction of the threshold voltage has resulted in loss of performance.

At present, amber is the only color available in a commercial TFEL display, however, recent work on zinc sulphide and other sulphides have shown that more efficient primary phosphors are possible. Several multi-color and full-color displays have been built in company laboratories but it appears that commercial full-color TFEL panels are still several years away.

Despite the existence of some technical difficulties, it is apparent that thin-film electroluminescent large-panel displays will become a serious threat to the traditional cathode ray tube at some time in the near future.

# FIBER OPTICS

## HISTORY OF FIBER OPTICS

Despite the fact that the modern science of fiber optics is considered to be relatively new, the principle used for this technology dates back to the fifteenth century. It was then that Venetian glass blowers recognized that light can follow a curved path when applied to one end of a bent, transparent or translucent glass rod or filament.

There was no major technical interest in this principle until 1870, when a presentation was made to the British Royal Society in London, illustrating that light could be conducted along a stream of water from a hole in the side of a tank. The presentation demonstrated the concept of producing total internal reflection of light within the boundaries of a transparent or translucent medium (in this case, water) and then have the light emerge at the other end of the medium.

In 1880, just four years after the invention of the telephone, Alexander Graham Bell experimented with the possibility of transmitting speech on a beam of light in air to a device he called the photophone. Unfortunately, it did not result in any significant technical breakthrough.

From 1910 through the 1930s, some isolated studies were done using the principle of light transmission in a translucent medium. It was not until the 1950s, that active research on this technology resulted in the development of a flexible device called a *fiberscope*, now widely being used in medicine as a diagnostic tool. It was during this period that a practical coated glass fiber was invented and the term "fiber optics" was coined. In 1967, researchers at the Standard Telecommunications Labs in England proposed that this technology could form the basis for a new communications medium.

In 1970, 100 years after the presentation at the British Royal Society, a major breakthrough occurred when the Corning Glass Works in New York developed a glass fiber that had relatively low losses – 10 decibel (db) per kilometer or losses of 70%. By today's standards, these losses would be considered very high. This accomplishment was the catalyst that stimulated new research programs and launched the fiber optic industry. Several existing companies and some new ones began the manufacture of glass and plastic fiber cable and required accessories.

Typically, the glass fiber optic cables used in modern telecommunications systems lose 1 db (less than 10%) of light per kilometer. Plastic fiber optics, used for short distances in display and light probe applications, have greater losses.

Although a plastic fiber optic cable has higher losses than glass, it is lighter, more rugged, can be bent at right angles without cracking and is easier to handle than glass.

There are two distinct causes of light losses – *absorption* and *scattering*. The main cause of absorption is the presence of some impurities – iron, copper, cobalt, vanadium and chromium; they have always been the source of contamination in high quality optical glass. For optical fibers, even one or two parts per billion of these metallic impurities are considered unacceptable. To produce "extremely pure" glass, a manufacturing process called *vapor deposition* has been used, providing the key to the advancement of fiber optic communications.

Scattering, the other source of light loss, is a change in the light wave caused by a change in the density of the medium. Small irregularities, the result of temperature variations in the mix as the glass solidifies, are very common in glass, but, to a great extent, can be controlled in the production process. Scattering losses are also related to the wavelength of light. At or near the infrared frequencies, light is more readily transmitted through glass fibers than light at other sections of the light spectrum. For this reason, light with wavelengths between 600 and 1600 nanometers is used for communicating through fiber optic cables.

## PRINCIPLE OF FIBER OPTICS

### STEP-INDEX FIBER

The simplest form of optical fiber is the *step-index* fiber and is based on the principle of total internal reflection within a single strand of fiber or rod. When a light beam is applied at one end, the generated light will stay within the boundaries of the medium and emerge at the other end.

The core of the fiber optic strand, approximately .01 to .05 inches in diameter, is made of either very pure glass (silica) or plastic. A *cladding*, a very thin glass or plastic material of a lesser density than the core, is applied to its outer surface. When a beam of light at a suitably low angle strikes the boundary between the two materials, the light is reflected back into the denser core. In this manner, light zig-zags through the core, reflecting off one side and then the other of the core-cladding interface, with the incident angle equal to the emerging angle. See Figure 147.

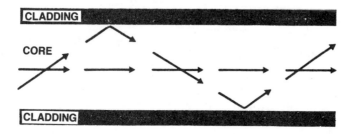

Step-index Fiber Optic Strand
**Figure 147**

A major problem with this technique of light transmission is that the transit time of the light beam that follows the zig-zag path is greater than that of the light beam traveling straight though the fiber. This means that over an extended distance, the sharpness of the light signal is significantly reduced, or distorted, limiting the system's capability.

GRADED-INDEX FIBER
The *graded-index* fiber provides less signal distortion. The density of the core changes gradually – high density at the center to low density at the core-cladding interface.

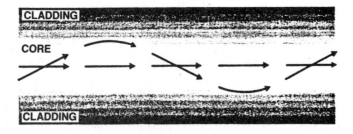

Graded-index Fiber Optic Strand
**Figure 148**

Instead of sharply zig-zagging, the light beams travel in smooth, curving paths, turning gradually back as they encounter the ever decreasing density on their way to the cladding. See Figure 148.

Although these curving beams travel nearly the same distance as they do in the step-index fiber, a larger portion of their travel is in the low-density, faster-traveling region. Transit time is therefore nearly that of the light beams that move in a straight line along the fiber's long axis.

MONO-MODE FIBER
In a *mono-mode* fiber, the diameter of the fiber core is made slightly larger than the wavelength of the transmitted light. (The diameter of the mono-mode core is generally between 2000 and 8000 nanometers.) All light beams travel in a straight path, arriving at the end of the fiber at the same time, providing a greatly increased information-carrying capacity with a minimum of distortion. See Figure 149.

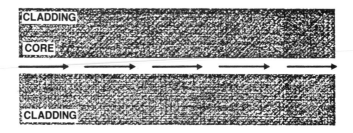

Mono-mode Fiber Optic Strand
**Figure 149**

Regardless of the particular mode being used, the light beam is merely the carrier of the desired signals being transmitted. The light must be *modulated*, or changed, with an audio, video or digital signal, by varying either the intensity of light (amplitude modulation) or the frequency of the light (frequency modulation). The modulated light beam will be received at the emerging end of the fiber strand and converted into electrical signals with the use of a photodetector. The audio, video or digital information is then removed from the converted carrier wave (demodulated) and the desired information amplified.

More light beams can be applied at the transmitting end of a fiber strand (each at a different angle of incidence for the step-index and graded-index fibers), with each light beam carrying a different modulated signal. The first commercial fiber optic strand was capable of carrying 67 different light beams, each independent of the other.

If only 10 strands were assembled into a single fiber optic cable, with each strand capable of carrying 67 light beams (channels), a total of 670 separate and non-interfering channels of communication is possible. Some cable configurations are shown in Figure 150.

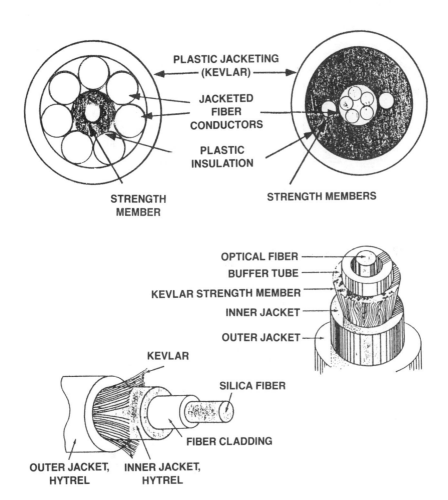

Cross-section of Typical Fiber Optic Cables
**Figure 150**

With more advanced fiber optic cables presently being manufactured, about ten times more channels of communication in one cable is now possible. With the present availability of precise modern optical systems used to detect each individual light channel, fiber optic technology is offering countless communications capabilities compared with copper cable systems.

To provide amplification of an attenuated light signal, it is necessary to convert the light into electrical signals with the use of photodetecting circuitry and amplifiers. The amplified electrical signal can then be used to energize a laser beam or infrared LED for use as the light source to a connecting fiber cable. A block diagram of a typical transmission system is shown in Figure 151.

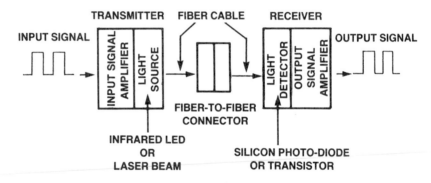

Block Diagram of a Typical Fiber Optic Transmission System
**Figure 151**

## FIBER OPTIC CABLE CHARACTERISTICS

The attributes of fiber optic technology are not limited to its capability of providing a multitude of communications channels. Although this characteristic, by itself, makes this technology very attractive, it is not the most significant of its many advantageous features.

Since light, not electricity, is being transmitted, the following features make the technology very effective:

● There are no short circuit possibilities.

● There is no spark or fire hazard.

- Fiber optics offer total immunity from radio frequency interference (RFI).

- Crosstalk between channels does not exist.

- Since there are no electrical signals to be radiated or magnetically induced, there is total transmission security.

- Cables presently in use can carry light information over a distance of about 100 kilometers (60 miles) before repeater stations are required to boost the transmitted light signal to usable levels.

- Fiber optic cable has a wide operating temperature range -
  Glass: $-60°C$ to $+200°C$
  Plastic: $-40°C$ to $+80°C$

- Although the initial cost of fiber optic cable is higher than copper cable, little or no maintenance is required. An evelution of total system cost, including maintenance, makes the use of fiber optics highly competitive with copper.

- The wide difference in frequency capability between a fiber optic cable and the copper wire used for conventional telecommunications systems is perhaps the most important feature of this technology. An ordinary copper line has an effective frequency capability of about 5000 Hertz. Only voice frequencies can be carried efficiently over telephone lines.

In comparison, the frequency capability of a fiber optic cable is in excess of 1000 Megahertz. This means that it is capable of handling audio, video signals (including cable television) and high speed digital information.


## FIBER OPTIC ADVANTAGES SUMMARY

TELECOMMUNICATIONS

- Extremely high cable capacity

- Improved performance – very high frequency capability

- Little or no cable maintenance costs

- Low cable losses – light can be handled over extremely long distances without the use of repeater stations

- No short circuit possibilities in fiber optic cable

- Total transmission security – no electrically detectable signals are radiated

- No crosstalk

## AEROSPACE, AIRFRAME, AVIONICS AND NAVIGATION

- Fiber optic cable weighs less and uses less space than copper cable providing the same monitoring channels

- Little or no cable maintenance costs

- Wide operating temperature range – glass fiber optic cable will operate from $-60°C$ to $+200°C$

## INDUSTRIAL CONTROL

- Immunity from RFI for computer-controlled systems that are being used to turn heavy equipment ON and OFF.

## SHIPBOARD

- No spark or fire hazard in fiber optic cable

It is quite obvious that as this exciting technology develops, the products that result will proliferate throughout all industries while new and dramatic innovations in the use of fiber optics, in one form or another, will be achieved.

# ANSWERS TO REINFORCEMENT EXERCISES

# CHAPTER TEN – LIGHT EMITTING DIODES (LED)

TRUE and FALSE:

1. True

2. False – The LED will emit light when a forward bias voltage (positive to the anode) is applied to the device.

3. True

4. True

5. True

6. False – Once the chip is committed to a specific substrate material and dopant concentration, it will always emit the same color. An increase in current will result in an increase in light intensity.

7. False – In addition to providing an enclosure and protection for the LED chip, the epoxy encapsulation serves to enhance the diffused appearance of the light.

8. False – The LED peak inverse voltage (PIV) rating is between 3 to 25 volts. The silicon diode PIV rating can range between 100 and 1000 volts.

9. True

10. False – The radiation pattern, provided on the data sheet, defines the luminous intensity of the LED when viewed at an off-axis angle. This information is required to properly specify the optical characteristics of the device.

11. True

12. True

13. False – High-efficiency, high-intensity red, yellow and green LEDs made from gallium phosphide material provide good readability in direct sunlight.

14. False – Light emitting diodes are more resistant than analog meters to mechanical shock and vibration.

15. False – The purpose of the resistor is to determine and limit the current in the circuit. Without the resistor, the current through the LED would be too high, resulting in destruction of the device.

16. True

17. False – In many applications, only moderately accurate visual indications of temperature, barometric pressure, speed of a moving vehicle and other information may be required. The electrical analog of these changes, in linear voltage form, can be used to drive either a linear or a circular array of LED lamps. If no need for a precision meter exists, this type of LED display can offer an effective, relatively inexpensive readout technique, with resistance to mechanical shock and vibration and high visibility in a low or moderately lit environmment.

# CHAPTER ELEVEN – LED DISPLAYS

TRUE and FALSE:

1. True

2. False – The 7-segment display is the most commonly used and least costly of all LED displays. It is used in calculators, digital clocks, digital meters, instrument panels and other similar display applications.

3. True

4. False – To obtain letters, "B" and "D", for a hexidecimal read-out and still use a standard 7-segment digital display, two additional small segments must be provided. With the use of these two small segments, the "B" can be distinguished from the "8" and the "D" from the "0". This display is often referred to as a 9-segment digital display.

5. False – For a 12-hour clock display, a colon (:) is used between the 1½ digit "hour" and the 2-digit "minute" display and is called a "3½ digit display with colon". The ½ digit is often referred to as the overflow digit.

6. True

7. True

8. False – The direct D.C. drive technique for LED displays is more efficient when there are fewer than four digits used for the display.

9. False – Since some of the drive circuitry is shared for all the display digits, multiplexing uses less total power and fewer parts than a direct D.C. drive system, reducing overall system costs. Multiplexing is generally preferred when there are four or more digits in the display.

10. True

11. False – To produce a flicker-free display, the strobing rate should be at least 100 Hz, however, a strobing frequency of 1 kilohertz or higher will provide optimum overall performance for LED displays.

12. False – Both single-digit and multiple-digit displays are available with decode/driver circuits provided as an integrated circuit and assembled on the same substrate as an *onboard integrated circuit.* (OBIC). Although this approach is more expensive than one using a display that does not incorporate an OBIC, the advantages can offset the higher initial cost.

13. False – Although human factors are difficult to evaluate quantitatively, they determine the size, shape and color of a display and are probably the most important considerations in the usability and success of a display.

14. True

15. True

16. False – The human eye is most sensitive to the wavelengths in the yellow-green area of the color spectrum.

17. True

# CHAPTER TWELVE – OTHER DISPLAY TECHNOLOGIES

TRUE and FALSE:

1. True

2. True

3. False – Because the screen printing technique that is used is a relatively simple process, any desired font or pattern can be easily screened on the glass. This includes bar graphs, in any desired form or size, both linear and circular, text of any nature or language and special symbols or graphics.

4. True

5. False – A high voltage is required to ionize a gas discharge display. Typical values are between 150 and 300 volts.

6. False – A green display can be produced by coating the front glass with an appropriate phosphor to absorb the ultraviolet energy being emitted by the neon gas, thereby releasing secondary light photons in the green region of the visible light spectrum. With the use of a filter, a red or amber display is possible.

7. True

8. False – Vacuum fluorescent panel displays are very readable in high ambient light conditions and will not wash out when exposed to the direct rays of the sun.

9. True

10. False – The most common display color is bluish-green, however, with appropriate filters, red, orange or yellow can also be obtained.

11. True

12 .True

13. True

14. False – Unlike all other displays, the LCD depends on ambient light to produce a visible display. This limitation is frequently overcome by the addition of an external light source (behind the display) that can be switched ON or OFF as desired. The display is most brilliant when exposed to the direct rays of the sun.

15. True

16. True

17. False – The voltage applied to the elements of a liquid crystal display must be in the form of A.C. If D.C. voltage is used, defects in the cell will eventually result.

18. False – Typical horizontal viewing angles range between 70° to 95°, with vertical viewing angles at about 60°. Several packaging techniques have been used to increase the horizontal viewing angle to about 120°; generally, the viewing angle for LCDs is much narrower than that of all other display technologies.

19. True

# CHAPTER THIRTEEN – OPTO-COUPLERS (OPTO-ISOLATORS)

TRUE and FALSE:

1. True

2. False – An opto-coupler can be used to prevent electrical noise or other voltage transients that may exist in one section of a circuit from interfering with a second section of the circuit, when the two sections are using a common circuit reference. Noise or voltage transients can be caused by a poorly designed common reference, creating a "ground loop" condition. This problem can easily be eliminated with the use of an opto-coupler.

3. True

4. False – The spectral response of silicon produces peaking at infrared frequencies, making silicon the most suitable element for use as photodetector material when an infrared LED is used as the emitter section.

5. True

6. False – The current transfer ratio (CTR) of a photodiode output opto-coupler is extremely low – about 0.10% to 0.15%.

7. True

8. False – The opto-coupler with a photodiode output has a very linear transfer characteristic at small signal levels. A small change in input current will produce a linearly proportionate change in output current. This characteristic of the opto-coupler makes it possible to couple either low-level analog signals (or small D.C. voltage variations) with little distortion.

9. True

10. True

11. False – Although a photo SCR is normally used to switch A.C. in the load, there may be a situation where it would be desirable to have the photo SCR operated with D.C. as the supply voltage. For example, a security or fire alarm system may require reset (using a SPST, momentary switch in the D.C. voltage supply circuit) of a latched-ON photo SCR output circuit to return the system to a standby condition.

12. True

13. False – Typical values of isolation voltage range between 500 and 6000 volts.

14. True

# CHAPTER FOURTEEN – SOLID-STATE RELAYS (SSR)

TRUE and FALSE:

1. True

2. False – Although the early solid-state relays were only able to switch an A.C. voltage at its output, newer SSRs, using power bipolar transistors and MOSFETs, can switch a D.C. voltage as well. The input (control) voltage to an SSR can be either D.C.(between 3 to 35 volts) or A.C. (between 90 to 250 volts at a frequency of 47 to 70 Hertz).

3. True

4. False – The opposite is true. At the moment the A.C. supply voltage is at a value other than zero, the trigger's output will be zero. When zero A.C. supply voltage is sensed, the Schmitt trigger will provide an output current.

5. False – Reducing the TRIAC gate current to zero does not turn the TRIAC OFF if the A.C. supply voltage is at a value other than zero. When the A.C. supply voltage crosses the zero level point, the TRIAC turns OFF. Since there is no current flowing in its gate circuit, the TRIAC stays OFF and at this point, the SSR is turned OFF.

6. True

7. True

8. True – State-of-the-art SSRs can switch 400 volts and 40 amperes using power MOSFETs in the output section.

9. True

10. False – Quiet operation is a secondary benefit of an SSR, unless the device is used in an environment where no sound at all is permitted. The other advantages of the SSR are generally much more important.

11. False – The opposite is true. Most electromechanical relays have the capability of multiple-pole, multiple-throw contact configurations. Until very recently, the solid-state relay was capable of only offering a single-pole, single-throw contact configuration, but now, SSRs are also being manufactured with double-pole-double-throw contact configurations.

12. False – Despite the inherent zero-crossing time delay of the SSR, switching action is faster than slower, "run-of-mill" electromechanical relays that usually take hundreds of milliseconds to switch. The longest delay incurred because of the zero-crossing characteristic is 8.3 milliseconds (when used with a 60 Hertz voltage supply).

13. False – SSRs have inherently higher resistance across their "contacts" than electromechanical relays. Because of this condition, the SSR generates more heat ($I^2R$), particularly at higher currents. Since the SSR dissipates higher power, it is not as temperature-tolerant as an electromechanical relay. Since the components of the SSR are more temperature-sensitive, some provision must be made for cooling. Adding cooling components, however, add more cost to the overall system.

14. True

# GLOSSARIES

### POPULAR OPTOELECTRONIC TERMS

### OPTOELECTRONIC SYMBOLS

# GLOSSARY OF POPULAR OPTOELECTRONIC TERMS

**ABSORPTION** – Losses in an optical fiber caused by the presence of some impurities – iron, copper, cobalt, vanadium and chromium. For optical fibers, even one or two parts per billion of these metallic impurities are considered unacceptable.

**ACCEPTANCE PATTERN** – The curve of total transmitted light plotted against the incident angle.

**ALPHANUMERIC DISPLAY** – A display used for numbers and letters.

**ANALOG DISPLAY** – A visual readout that provides a continuous illuminated movement along a predetermined calibrated path to provide an indication of a desired measurement.

**ANALOG METER** – A visual readout that provides a continuous numbered scale along which a pointer moves.

**CLADDING** – A very thin glass or plastic material applied to the outer surface of the core of a fiber optic strand having less density than the core. When a beam of light at a suitably low angle strikes the boundary between the two materials, the light is reflected back into the denser core.

**CONNECTOR INSERTION LOSS** – The light loss, expressed in decibels (db), caused by the insertion of a mating connector between two sections of a fiber optic transmission system.

**CONTRAST** – The difference in light output between the illuminated and nonilluminated areas of a display.

**CORE** – The inner section of a fiber optic strand made of either pure glass (silica) or plastic through which light is guided. Typically, it is approximately .01 to .05 inches in diameter.

**CURRENT TRANSFER RATIO (CTR)** – The ratio of an opto-coupler's output current to its input current under specified conditions.

**DARK CURRENT** – The output current of a photodetector with no light at its input.

**DECODER/DRIVER** – A circuit that converts digital information into voltages, then amplifies the voltages to an adequate level to energize the elements of a display.

**DETECTOR** – A component that converts optical energy (light) into electricity. See PHOTODETECTOR.

**DIGITAL DISPLAY** – A visual readout of alphanumerics (letters and numbers) as opposed to analog or linear meter indications.

**DOMINANT WAVELENGTH** – The wavelength that is a quantitative measure of the color of light as perceived by the human eye.

**DOT-MATRIX DISPLAY** – A formation of dots arranged in an array of rows and columns, capable of being individually illuminated to produce a variety of alphanumeric characters and graphics.

**DYNAMIC SCATTERING LCD** – One technique used in liquid-crystal displays that causes ambient light to be scattered when the LCD segments are properly energized, producing a visual readout.

**ELECTROLUMINESCENT (EL) DISPLAY** – A display having segments or dots of transparent conductive electrodes separated by a thin dielectric containing a luminescent phosphor. The application of an A.C. voltage across opposing electrodes will cause the dielectric to glow with a characteristic bluish-green light.

**EMITTER** – The light source used in a fiber optic system.

**FIBER OPTICS** – The technology of transmitting and guiding of optical radiation (light) along optical conductors.

**FIBER OPTIC CABLE** – A bundle of many optical fibers within one conducting cable. In addition to the optical fibers and its cladding, the cable can consist of plastic strength members, an inner plastic jacket and an outer plastic jacket, all providing protection to the fiber strands.

**FIBER OPTIC CONNECTOR** – A quick connect/disconnect assembly used to interconnect: a light source to a fiber optic cable; a fiber optic cable to a fiber optic cable; a fiber optic cable to a light detector.

**FIBER OPTIC COUPLER** – A mechanical component that interconnects a number of fiber optic cables to provide a bidirectional system by mixing and splitting all light signals within the cable.

**FIBER OPTIC SPLICE** – A nonseparable junction joining two optical fibers. The connection is made with the use of an epoxy adhesive or by thermal bonding.

**FLUORESCENT DISPLAY** – A vacuum tube display with a phosphor-coated screen that glows with a characteristic bluish-green light when bombarded with electrons from a heated filament.

**GALLIUM ARSENIDE (GaAs)** – A compound of gallium and arsenic used in the production of LED chips.

**GALLIUM ARSENIDE PHOSPHIDE (GaAsP)** – A compound of gallium, arsenic and phosphorus used in the production of LED chips.

**GALLIUM PHOSPHIDE (GaP)** – A compound of gallium and phosphorus used in the production of LED chips.

**GAS DISCHARGE DISPLAY** – A display in which patterned segments or dots, containing an inert gas, are under the surface of a glass screen. The display is illuminated by the application of a sufficiently high voltage to ionize the gas and emit a light on the screen. The light is generally characterized by a bluish-green color. Also called a PLANAR or PLASMA display.

**GRADED-INDEX FIBER** – An optical fiber whose gradually changing density provides a high density at its center and a low density at the core-cladding interface. Instead of reflected light sharply zig-zagging, the reflected light in the fiber travels in a smooth, curving path, reducing distortion in the fiber.

**GROUND LOOP NOISE** – An undesirable voltage generated in the common reference of a relatively low-level signal circuit by magnetic fields or by the return or reference currents produced by relatively high-power circuits nominally connected to the same circuit reference (ground). This is a potentially detrimental condition, generally caused by poor circuit layout.

**ILLUMINATED LEGEND DISPLAY** – A display in which information is imprinted on a translucent surface. This display is generally illuminated by backlighting.

**INFRARED RADIATION** – Radiation in the electromagnetic wavelength region situated between 7500 and 10,000 nanometers.

**IONIZATION** – The breakdown of an inert gas when a high voltage of sufficient amplitude is applied across it.

**IRED** – An infrared emitting diode. See LIGHT EMITTING DIODE.

**LEAST SIGNIFICANT DIGIT** – The numeral at the right-hand end of a digital numerical display.

**LED ARRAY** – A number of independent LED chips in a package.

**LIGHT CURRENT** – The current flowing through a photodetector when light is applied to its input.

**LIGHT EMITTING DIODE (LED)** – A semiconductor diode that emits essentially monochromatic (single color) light when forward biased. The emitted light can be red, yellow, orange, green or nonvisible infrared.

**LIGHT REPEATER** – An electronic circuit that converts a light signal to an electrical signal, amplifies the electrical signal and then changes the electrical signal back to an amplified version of the original light signal. Essentially, it is a light amplifying circuit when applied to a fiber optic system.

**LINEAR READING** – An equal change in visual indication for a given change in the quantity under measurement. The term LINEAR is generally interchangeable with the term ANALOG.

**LIQUID-CRYSTAL DISPLAY (LCD)** – A display having conductive segments or dots deposited on the inside surfaces of two transparent glass plates separated by a crystal in liquid form. When energized with an A.C. voltage in the presence of light, the selected segments will provide a gray or black tone readout.

**LUMINANCE CONTRAST** – The observed brightness of a light emitting diode display element compared with the brightness of the surrounding field that is an integral part of the device.

**MONO-MODE FIBER** – An optical fiber process where the diameter of the core is slightly larger than the wavelength of the transmitted light. All light beams travel in a straight path through the fiber, providing a greatly increased information-carrying capacity with a minimum of distortion when compared with the step-index and graded-index fibers.

**MOST SIGNIFICANT DIGIT** – The numeral at the left-hand end of a digital numerical display.

**MULTIPLEXING** – A display drive technique used with multiple-digit (four or more) displays. Selected digits are pulsed at a high peak current and low duty cycle so that the human eye cannot determine when the digits are being turned ON or OFF. As a result, the total power required for the display is reduced, with fewer parts used, since some of the drive circuitry is shared by all the display digits. Also called STROBING.

**NEMATIC LIQUID** – A liquid-crystal material used as the display medium in a liquid-crystal display. A nematic liquid is characterized by having its crystalline structure, or lattice, totally aligned when no energizing voltage is applied, resulting in a transparent liquid. See LIQUID-CRYSTAL DISPLAY.

**NUMERICAL APERTURE (NA)** – The characteristic of an optic fiber which defines its acceptance of impinging light. The "degree of openness", "light gathering ability" and "angular acceptance" are other terms describing this characteristic.

**OPTO-COUPLER (OPTO-ISOLA-TOR)** – A component capable of optically transferring an electrical signal between two circuits and, at the same time, electrically isolating these circuits from each other. It consists of an infrared LED emitting section at its input and a silicon photodetector at its output with other circuitry sometimes included as part of the device.

**OPTOELECTRONICS** – The technology that merges the sciences of electronics and optics.

**PEAK WAVELENGTH** – The wavelength at the peak of the radiated spectrum of a light emitter.

**PHOTO DARLINGTON** – A pair of bipolar transistors connected in a Darlington configuration to provide very high current gain and used as the photodetector section of an opto-coupler.

**PHOTODETECTOR** – A device capable of sensing light and converting it into electricity.

**PHOTODIODE** – A silicon diode used as a photodetector.

**PHOTO SCR** – A silicon controlled rectifier used to act as the photodetector section of an opto-coupler.

**PHOTOTRANSISTOR** – A bipolar transistor used as a photodetector. It provides a current at its output that is proportional to the light intensity at its input. The low-level light current generated at its input is amplified by the current gain of the transistor.

**PLANAR or PLASMA DISPLAY** – See GAS DISCHARGE DISPLAY.

**POINT SOURCE** – A light source with a maximum dimension less than one-tenth the distance between source and detector.

**POLARIZER or POLARIZED FIL-TER** – A transparent film that allows light to pass through it in one plane and blocks light in other planes.

**SCATTERING** – A change in the light wave passing through an optical fiber caused by an impurity or change of density in the fiber. This effect produces losses in the fiber.

**SEGMENTED DISPLAY** – A rectangular pattern of segments arranged to provide an alphanumeric display.

**SPLICE** – A nonseparable junction joining two fiber optic conductors. It is made by fusing with heat or an epoxy adhesive.

**STEP-INDEX FIBER** – The simplest form of optical fiber that is based on the principle of total internal reflection within the fiber strand or rod. When a light beam is applied at one end, the light will bounce through the fiber in a zig-zag manner and emerge at the other end.

**STROBING** – See MULTIPLEXING.

**SUPERTWISTED BIFRINGENCE EFFECT (SBE)** – An advanced approach to liquid-crystal displays which offers a distinct improvement in contrast (particularly in ambient low-level light) over the older nematic field-effect LCD.

**THIN-FILM ELECTROLUMINES-CENT TECHNOLOGY (TFEL)** – An advanced approach to the production of large-panel fluorescent displays that allow them to compete with other display technologies in data-display and graphics-display applications.

**VAPOR DEPOSITION** – A step in the manufacture of optical glass fiber to reduce the inherent metallic impurities in the material.

# GLOSSARY OF OPTOELECTRONIC SYMBOLS

# GLOSSARY OF OPTOELECTRONIC SYMBOLS

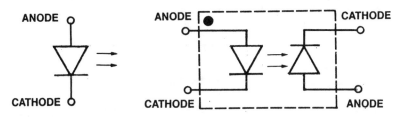

Light Emitting Diode
**(LED)**

Opto-coupler with Photodiode Output

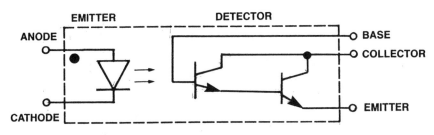

Opto-coupler with Phototransistor Output

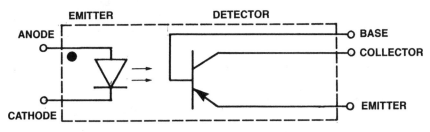

Opto-coupler with Photo Darlington Output

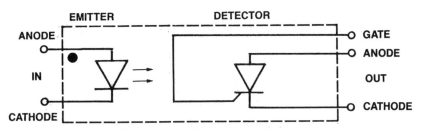

Opto-coupler with Photo SCR Output

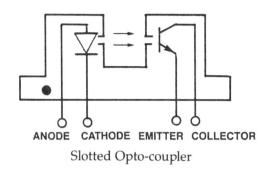

ANODE  CATHODE  EMITTER  COLLECTOR

Slotted Opto-coupler

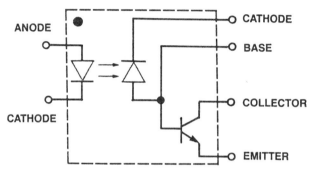

Opto-coupler with Photodiode and Transistor Output Section

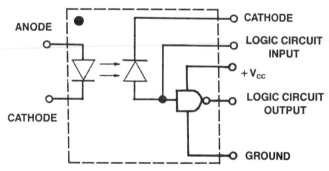

Opto-coupler with Photodiode and Logic Circuit Output Section

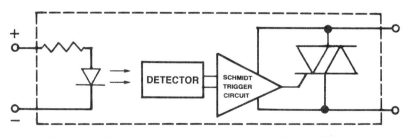

Opto-coupled, Zero-crossing Solid-state Relay (SSR)

## DIGITAL DISPLAY FORMATS

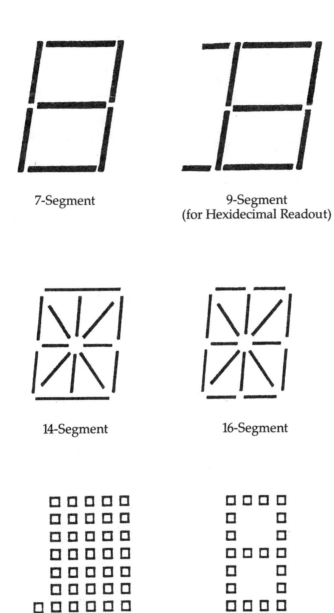

7-Segment

9-Segment
(for Hexidecimal Readout)

14-Segment

16-Segment

5 × 7 Dot-matrix
with left hand decimal

Modified 4 × 7
Dot-matrix

# REFERENCES

MODERN DICTIONARY OF ELECTRONICS – Rudolf Graf
– Howard W. Sams, Inc.

ELECTRONIC BUYERS' HANDBOOK & DIRECTORY – 1986
– CMP Publications, Inc.

ELECTRONIC ENGINEERING TIMES
– CMP Publications, Inc.

ELECTRONIC ENGINEERS MASTER (EEM) – 1986/1987
– Hearst Business Communications, Inc.

PHYSICS – Hausman and Slack
– D. Van Nostrand Company

SEMICONDUCTOR POWER CIRCUITS HANDBOOK – 1968
– Motorola Inc. – Semiconductor Products Division

SWITCHING TRANSISTOR HANDBOOK – 1970
– Motorola Inc. – Semiconductor Products Division

GENERAL TRANSISTOR MANUAL – 1960
– General Electric Company

SILICON CONTROLLED RECTIFIER DESIGNERS' HANDBOOK – 1963
– Westinghouse Electric Company

SILICONIX SEMICONDUCTOR DEVICES
– Siliconix Incorporated

THE RELIABILTY HANDBOOK – 1979
– National Semiconductor Corp.

OPTOELECTRONICS APPLICATIONS MANUAL – 1977
– McGraw-Hill Book Company

SOLID-STATE PRODUCTS DATA BOOK – 1984
– Teledyne Solid-state Products

FLUORESCENT INDICATOR APPLICATION NOTES
– NEC Electronics Inc.

# INDEX

## A

A.C. switch, 73
absorption, 314
acceptance quality level (AQL), 165, 171
alphanumerics, 210, 235, 255
aluminum, 13, 140
amplification, 3
amplitude modulation of light, 316
analog amplifier, 199
analog meter, 229
analog switch, 155, 199
Angstrom unit, 226
anode, 30, 36, 74
answers to reinforcement exercises -
    discrete semiconductors, 185
    optoelectronics, 321
arsenic, 13, 14
ASCII, 239
automatic frequency control (AFC), 55
avalanche region, 45

## B

backlighting, 267, 270, 272, 308, 309
bar graph, 210, 256
Bardeen, John, 4
base, 103
    current, 104
Bell Telephone Labs, 4
Bell, Alexander Graham, 313
beryllium oxide insulator, 98
beta (a), 107, 115
bipolar transistor, 10, 103, 296
    applications, 118
    characteristics and trade-offs, 121
    specifications, 112
    subscript notations, 123
boot-strap circuit, 97
boron, 13, 28, 132, 140
Brattain, Walter, 4
bridge rectifier, 10, 61
British Royal Society, 313

# H

# I

# P

P-channel FET, 131
P-region, 29, 35, 36
passivation, 27
passivation layer, 29
peak inverse (reverse) voltage, 40
peak wavelength, 226, 249
phosphor, 258
phosphor-coated screen, 261, 262
phosphorus, 13, 14, 28, 132, 140
photo Darlington, 282, 284, 294
photo SCR, 283
photo-voltaic effect, 54, 281, 296
photodetector, 279, 284, 294, 317
photodiode, 10, 54, 279, 281
photomasking, 27
photon, 222, 258
photophone, 313
photoresist, 27, 28
phototransistor, 279, 282, 284, 294
picture element (pixel), 308, 311
pinch-off voltage, 135
planar display, 256
plasma display, 212, 256
PN junction, 29, 36
PNP transistor, 103, 104
point source, 340
polarizer, 307, 308, 340
polycrystalline silicon, 24
polymerize, 28
power diode geometry, 28
power burn-in, 169
power derating, 41
power MOSFET, 177, 297
    packaging, 180
power semiconductors, 11

# Q

qualified parts list (QPL), 167
quality conformance testing, 167
quality control program, 170
Quality Specifications – MIL-Q-9858, 170
quiescent collector current, 108

# R

radiation pattern, 227
radio frequency amplifiers, 13

# T